Making Solar Pay

Making Solar Pay

A Financial Guide to Solar Electric Systems

Matt McNearney

Mountain Edge Publishing

Copyright © 2017, 2018 Matt McNearney

All rights reserved. No part of this book may be reproduced in any form or by any means without expressed permission in writing.

ISBN: 978-1539327264

Acknowledgements

The author and publisher wish to extend a special acknowledgement with gratitude to **Steve Risley**, whose insights and detailed reviews helped make this a better book.

In addition, the author and publisher thank the following people for their contributions to the manuscript:

- Shawn Ransom
- Michael DeRosia
- Jeff Fleischman
- Oz Pfenninger

Any errors within the book are the author's, not theirs.

Contents

	Introduction	9
1.	Getting Started: Estimating Electrical Usage	11
2.	Types of PV Systems	21
3.	Incentives	41
4.	Buying Option	47
5.	Proposals	57
6.	Financial Analysis	69
7.	Installation	83
8.	Maintenance	89
9.	Tracking Performance	91
10.	Case Study: How One Homeowner Used This Book to Determine Whether Solar Makes Financial Sense	99
	Appendix A - Other Types of PV Systems	109
	Appendix B - Spreadsheet: Cash Option	117
	Appendix C - Spreadsheet: Solar Loan Option	129
	Appendix D - Spreadsheet: Mortgage Option	139
	Index	147

Introduction

Solar energy makes sense as a concept, but does it make sense financially? That question should be asked by anyone considering a solar electric system, and it is the driving force behind this book.

The book takes the homeowner's perspective. You don't need to know the ins and outs of electricity. And you certainly don't need to know the technical details of a photovoltaic system. I am content to say that all of this works by magic. Hey, sunlight hits some modules on my roof, and the lights go on in my house. Seems like magic to me.

The focus is financial. The book gives you the details and tools you need to make an informed decision about a solar electric system, including:

- Information that a solar company needs to create a custom proposal for you
- Basic system components so you can better understand those proposals
- Various incentives available to you
- Your buying options: using cash, borrowing the money, or leasing a system
- Comparing proposals from competing companies

- Conducting your own financial analysis of the best proposals
- Common installation problems and how to prepare for them
- Maintenance issues
- Tracking performance to ensure the system is operating effectively

What the book does not do is take a position in the holy war between renewable energy and fossil fuels. Renewable energy is the future, but for the time being, it is very clear that we need both. It doesn't make sense (economic or otherwise) to rely on one source of energy to the exclusion of others. Today's challenge is to build a diversified energy portfolio, one that enables us to adapt to whatever comes our way.

In building that portfolio, money spent on solar should be considered an investment. However, you should never overpay for any investment, no matter how important it is or how well it performs.

At some point solar energy will be cost effective. This might be the result of economies of scale as more people adopt it, better efficiencies from technical advances, or the inevitable price increases for fossil fuels. It will happen; it's just a matter of when.

For many, that time is now. Solar energy is already a good investment for them. This book will help you decide whether it is a good investment for you.

One
Getting Started: Estimating Electrical Usage

How much electricity do you use each year? A solar company is sure to ask that question, most likely within the first few minutes of your initial conversation. How you answer depends on your particular situation. There are three likely scenarios.

Scenario 1: No House, No Data

Obviously, you cannot provide historical data of your electrical usage if your house has not been built yet. The same applies if current data is no longer valid due to a "gut the house and start over" project or a significant add-on renovation. In these instances, you will need to provide the square footage of the new/renovated house and a copy of the construction plans that will be used for the work. The square footage will enable the bidding solar companies to estimate your future electrical usage and the system size needed to meet your needs. The plans will show whether you will have space to accommodate the optimal system and the best areas for locating it.

The example on the next page shows both the square footage and the finished design of the new roofing structure.

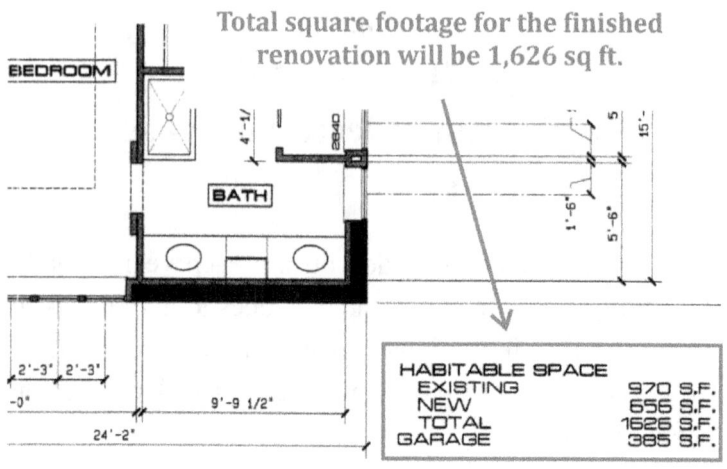

Total square footage for the finished renovation will be 1,626 sq ft.

HABITABLE SPACE	
EXISTING	970 S.F.
NEW	656 S.F.
TOTAL	1626 S.F.
GARAGE	385 S.F.

The exterior elevation shows possible locations to fit a solar electric system.

Scenario 2: New House, Limited Data

If you moved into a house recently (meaning less than twelve months ago), then you probably don't have enough historical data on electrical usage there. For brand-new construction, all you need to provide bidding solar companies is your square footage. They can estimate your likely usage from that, just as they would for a house that is currently under construction.

Estimating Electrical Usage

If the house had a previous occupant, there may be historical data, but you likely don't have it. Perhaps the utility company does. You should contact your utility to see if they have available details on past electrical usage. The other option is to simply provide bidding solar companies with the square footage (and any electric bills that you do have), and they can do the usual estimates.

In either case, a solar company can probably use online satellite imagery of your house to identify a good location for a solar electric system. (It's kind of scary what you can see with online satellite images. If you are curious, go to an online resource like Google Maps and enter your address. And the next time you are in your backyard, be sure to look up and wave.)

Scenario 3: Not a New House, Plenty of Data

The third scenario is when you have lived in your house for at least a year. For this, a solar company will want to see twelve months of electrical data, the reason being that your usage can vary significantly, not so much from month to month but certainly from one season to the next. By looking at data for an entire year, you can smooth out the peaks and valleys and get an accurate picture of your overall consumption.

Probably the easiest way to get the data is to log in to your utility company's website and see if you can access a report of your usage. The key number is kilowatt-hours or kWh. That is the amount of electricity used.

Making Solar Pay

YOUR PAST 12 MONTHS OF ENERGY USE AND CHARGES

Last Read Date	Billing Days	Avg Temp (°F)	Kilowatt Hours	Electricity Charges**	Carbon Footprint (lbs) #
April 14, 2015	29	51°	766	$ 89.66	1024.91
May 13, 2015	29	50°	754	$ 88.21	1008.85
June 12, 2015	30	58°	883	$ 109.99	1181.45
July 14, 2015	32	70°	1,009	$ 139.22	1350.04
August 13, 2015	30	73°	1,180	$ 160.84	1578.84
Sep			1,042	$ 139.38	1394.2
Oct			688	$ 82.84	920.54
Nov			588	$ 68.82	786.74
Dec			677	$ 78.01	905.83
Jan			763	$ 84.79	1020.89
Feb			636	$ 69.47	850.97
March 15, 2016	32	47°	688	$ 74.53	920.54
Totals			9,674	$ 1,185.76	12,943.80

Total kWh is the number to give to solar companies.

One nice thing about the example report is that it also provides the amounts that you paid for that electricity each month and for the year. You just need to take this a step further and divide the total cost by kilowatt-hours. This will give you the cost per kWh, which is a number that we will use in our financial analysis.

Last Read Date	Billing Days	Avg Temp (°F)	Kilowatt Hours	Electricity Charges**	Carbon Footprint (lbs) #
April 14, 2015	29	51°	766	$ 89.66	1024.91
May 13, 2015	29	50°	754	$ 88.21	1008.85
June 12, 2015	30	58°	883	$ 109.99	1181.45
July 14, 2015	32				
August 13, 2015	30				
September 14, 2015	32		$1,185.76 / 9,674 = 0.1226		
October 13, 2015	29				
November 10, 2015	28		$0.1226 per kWh		
December 11, 2015	31				
January 14, 2016	34	30°	763	$ 84.79	1020.89
February 12, 2016	29	36°	636	$ 69.47	850.97
March 15, 2016	32	47°	688	$ 74.53	920.54
Totals			9,674	$ 1,185.76	12,943.80

14

Estimating Electrical Usage

Without a report that provides these details, you will have to compile the data by hand. The best way is to start with the most current electric bill and then work backward to gather data for the previous twelve months. For each statement, you need your kilowatt-hour usage and amount you paid. Make sure you're using only the portion for the electrical service if your utility provides both electricity and natural gas. You don't want to use the total cost of the bill, as that would include the charges for gas as well.

For each month, you need kWh usage and total expense.

SERVICE ADDRESS	ACCOUNT NUMBER	DUE DATE
		09/02/2015
	STATEMENT DATE	AMOUNT DUE
	08/13/2015	$160.84

ELECTRICITY SERVICE DETAILS

Meter Reading Information

Read Dates: 07/14/15 - 08/13/15 (30 days)

Description	Current Reading	Previous Reading	Usage
Total Energy	63430 Actual	62250 Actual	1180 kWh

ELECTRICITY CHARGES RATE: R Residential General

Description	Usage Units	Rate	Charge
Service & Facility			$6.75
Summer Tier 1*	500 kWh	$0.046040	$23.02
Summer Tier 2*	680 kWh	$0.090000	$61.20
Trans Cost Adj	1180 kWh	$0.000630	$0.74
Elec Commodity Adj	1180 kWh	$0.027810	$32.82
Demand Side Mgmt Cost	668.67 kWh	$0.001220	$0.82
Demand Side Mgmt Cost	511.53 kWh	$0.001550	$0.79
Purch Cap Cost Adj	1180 kWh	$0.006500	$7.67
CACJA	1180 kWh	$0.003920	$4.63
Renew Energy Std Adj			$2.94
GRSA			$9.27
Subtotal			**$150.65**
Franchise Fee		3.00%	$4.52
Sales Tax			$5.67
Total			**$160.84**

15

You can download a spreadsheet to help organize the data by going to the Mountain Edge Publishing website (www.mepub.com). Simply enter the kWh usage and cost numbers for each month. The spreadsheet calculates the totals for both of those and the cost per kilowatt-hour. Again, the number to give the solar companies is total kilowatt-hours used.

	Electrical Usage and Cost		
	Past 12 Months--kWh Usage and Cost		
	Month	kWh used	Cost
1	March 2016	688	74.53
2	February 2016	636	69.47
3	January 2016	763	84.79
4	December 2015	677	78.01
5	November 2015	588	68.82
6	October 2015	688	82.84
7	September 2015	1,042	139.38
8	August 2015	1,180	160.84
9	July 2015	1,009	139.22
10	June 2015	883	109.99
11	May 2015	754	88.21
12	April 2015	766	89.66
	Totals	9,674	1,185.76
	Cost per kWh		0.1226

Total kWh is the number to give to solar companies.

It's possible that you may have access to more data than just the past twelve months. The utility's website might allow you to generate reports using data from two, three, or more years ago, or you might have several years of electric bills stored on your computer or stuffed in a drawer somewhere. By comparing the annual usage from several years, you can see if the most recent year's usage is a reasonable number to use. Should there be a large discrepancy, think back on possible reasons for that. Maybe an unusual year weather-wise caused you to run your air conditioner more than normal. Or perhaps recent home upgrades changed how much energy you use. The solar companies can help you analyze your usage to determine the best number to use for an estimate.

There is one last point to make about using data from past years. If you decide to calculate an average kWh-usage number using two or more years, you should still calculate the cost per kilowatt-hour using only the data from the most recent twelve months. The rates that utility companies charge for electricity tend to rise year over year. The most current rate is what we will use as the starting point for financial projections.

What's Your Goal?

Once you have your usage number, the next question you need to answer is,

What is your goal for your solar electric system?

Generally, there are two ways to go with this. One is to offset 100 percent of electrical usage. In other words, if you use 10,000 kilowatt-hours annually, then the goal is to have your solar electric system produce 10,000 kWh over the course of a year. Keep in mind that offsetting 100 percent of the electricity will not completely eliminate your electric bill. You will still have fixed costs, such as service and connection fees, but you should be able to eliminate most of the variable costs.

To identify what the variable costs are, look at your electric bill. These are the charges that are determined by a rate multiplier. You can also consider many of the other fees and taxes as variable charges since they are based on a percentage of the overall electric bill.

Variable Charges

ELECTRICITY CHARGES	RATE: R Residential General		
Description	Usage Units	Rate	Charge
Service & Facility			$6.75
Summer Tier 1*	500 kWh	$0.046040	$23.02
Summer Tier 2*	680 kWh	$0.090000	$61.20
Trans Cost Adj	1180 kWh	$0.000630	$0.74
Elec Commodity Adj	1180 kWh	$0.027810	$32.82
Demand Side Mgmt Cost	668.67 kWh	$0.001220	$0.82
Demand Side Mgmt Cost	511.53 kWh	$0.001550	$0.79
Purch Cap Cost Adj	1180 kWh	$0.006500	$7.67
CACJA	1180 kWh	$0.003920	$4.63
Renew Energy Std Adj			$2.94
GRSA			$9.27
Subtotal			$150.65
Franchise Fee		3.00%	$4.52
Sales Tax			$5.67
Total			$160.84

Charges based on percentages

A second option is to offset only a portion of your electrical usage. This is frequently done in locations where the utility has a tiered billing structure. With a tiered structure, the utility charges one rate for usage up to a certain level and then a higher rate for usage above that level. The net result is that the more electricity that you use, the higher the rate you pay.

Estimating Electrical Usage

This can get expensive, particularly in warmer climates, where running an air conditioner constantly can lead to very high per-kWh rates. You might also consider a partial offset if your home has certain solar restrictions, such as limited space available, the roof's orientation to the sun, and shade from nearby trees and structures.

The sample bill below has a tiered structure that kicks in during the summer months. During this time, you are charged one rate for the first 500 kWh and then a second rate for all kWh above 500. The tier 2 rate is almost twice that of tier 1. The goal for your solar electric system might be to generate enough electricity to ensure that you stay below the threshold for the higher rate.

ELECTRICITY CHARGES	RATE: R Residential General		
Description	Usage Units	Rate	Charge
Service & Facility			$6.75
Summer Tier 1*	500 kWh	$0.046040	$23.02
Summer Tier 2*	680 kWh	$0.090000	$61.20
Trans Cost Adj	1180 kWh	$0.000630	$0.74
Elec Commodity Adj	1180 kWh	$0.027810	$32.82
Demand Side Mgmt Cost	668.67 kWh	$0.001220	$0.82
Demand Side Mgmt Cost	511.53 kWh	$0.001550	$0.79
Purch Cap Cost Adj	1180 kWh	$0.006500	$7.67
CACJA	1180 kWh	$0.003920	$4.63
Renew Energy Std Adj			$2.94
GRSA			$9.27
Subtotal			$150.65
Franchise Fee		3.00%	$4.52
Sales Tax			$5.67
Total			$160.84

Obviously, the savings with a partial offset will not be as high as it would be if you offset 100 percent. Also obvious, you can accomplish a partial offset with a smaller system, so you will have a smaller overall investment. The Mountain Edge Publishing website has financial-analysis spreadsheets to help you determine which goal is best for you.

To summarize, you need to provide a solar company with two bits of information:

1. An estimate of your electrical usage
   - For new construction or significant renovation: square footage of completed house and project plans
   - For existing house but limited historical data: square footage or report of past usage at that house, if available
   - For current house with more than one year's worth of data: total number of kilowatt-hours that you used during the previous twelve months

2. What you hope to accomplish by installing a solar electric system, be it a 100 percent offset of your annual usage or a partial offset to reduce your electrical expense each month

The solar company will take that information and build a proposal for a customized PV system. But what exactly is that? All we've talked about so far are usage estimates and electric bills. We haven't spent any time on what makes up a solar electric system.

That is the topic of the next chapter.

Two
Types of PV Systems

A photovoltaic (PV) system is one that generates electrical power from the sun. The design of these systems can range from fairly simple to very complex. This chapter covers both ends of that spectrum.

Basic Components of a PV System

Any PV system starts with a photovoltaic **cell**, the smallest component that produces electric power when sunlight hits it. Solar manufacturers string a series of cells together to make a **module**, also referred to as a panel. A series of modules strung together is known as an **array**.

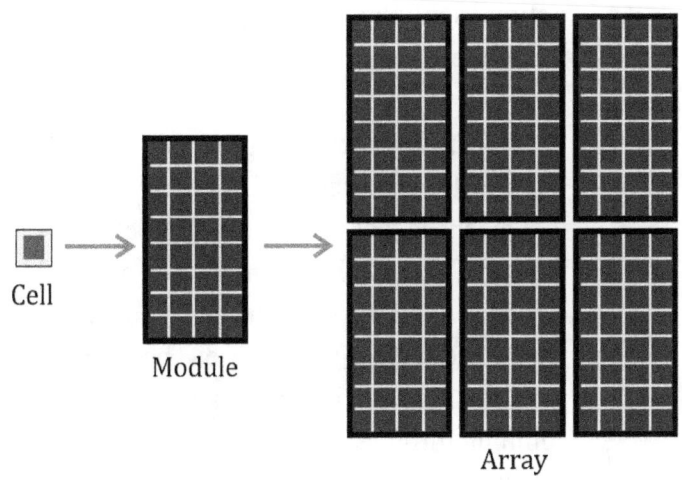

Each PV cell in a module has the capacity to produce a small amount of electric power. Adding together the potential production of all the cells gives you the maximum output for that module. Then adding the output of all the modules in an array determines the overall size of the system. For example, connecting eight modules that can generate 250 watts each would give you an array that can produce 2,000 watts of power, or 2 kilowatts. That would be referred to as a 2 kW system.

The list below presents the basic components of a PV system:

- **Array**: The size of the array is its maximum production rate. If that 2 kW system has 5 hours of optimal sunlight, then estimated production would be 2 kilowatts * 5 hours = 10 kilowatt-hours, or 10 kWh. The PV system will also generate electricity when there is not optimal sunlight; it just won't be at that maximum rate.

- **Inverter**: The electricity produced by a PV array is direct current or DC electricity. What this means is that the electrons knocked off the cells by sunlight are moving in one direction, flowing from the cells, down the wires, and toward your house. The electricity provided by the utility is created from spinning turbines, which results in the electrons moving in a wave. This is known as

Types of PV Systems

alternating current or AC electricity. In order for the electricity produced by your PV array to be compatible with the electricity used inside your house, it must be converted from DC electricity to AC electricity. That is the job of the inverter.

- **Electric service panel**: This is the box full of circuit breakers, which you already have. The service panel takes the electricity from a PV system or the grid and disperses it through the breakers to the various house loads (refrigerators, air conditioning units, lights, televisions, etc.).

- **Electric meter**: This is another item that you already have, although you will need a new one for a PV system. Your current electric meter works only in one direction, measuring the amount of electricity pulled from the utility grid. The new meter goes forward and backward. It keeps track of the electricity coming into your house from the utility as well as the excess electricity generated by your PV system that is sent out to the grid.

You can add additional components to a PV system, such as a battery, but those can make the system design more complicated and much more expensive, both to install and to maintain. We will get to a discussion of batteries, but there is an important concept that you need to know going into that: the basic residential PV system already has a battery-like component built in to it. It's known as the utility company's power grid.

Grid-Direct System

The most common type of PV system installed in the United States is the grid-direct system. That system consists of the four basic components that were just covered: a PV array, an inverter, an electrical service panel, and an electric meter. As the name implies, the system is connected directly to the utility company's electrical grid. This design leads to three possible scenarios, shown on the next few pages:

Scenario 1: PV System Supplies All Electricity

The PV system produces all electrical power needed by the house and sends excess electricity out to the grid. The steps for doing this are as follows:

1. The PV array generates DC electricity when sunlight hits it.

2. The DC electricity flows to the inverter, which converts it to AC electricity.

Types of PV Systems

3. The AC electricity flows to the electrical service panel from which it is dispersed to the various systems and equipment inside the house. These are usually referred to as loads.

4. Any AC electricity not used by the house flows out to the utility's power grid, passing through the electric meter, which keeps track of how much electricity was sent out.

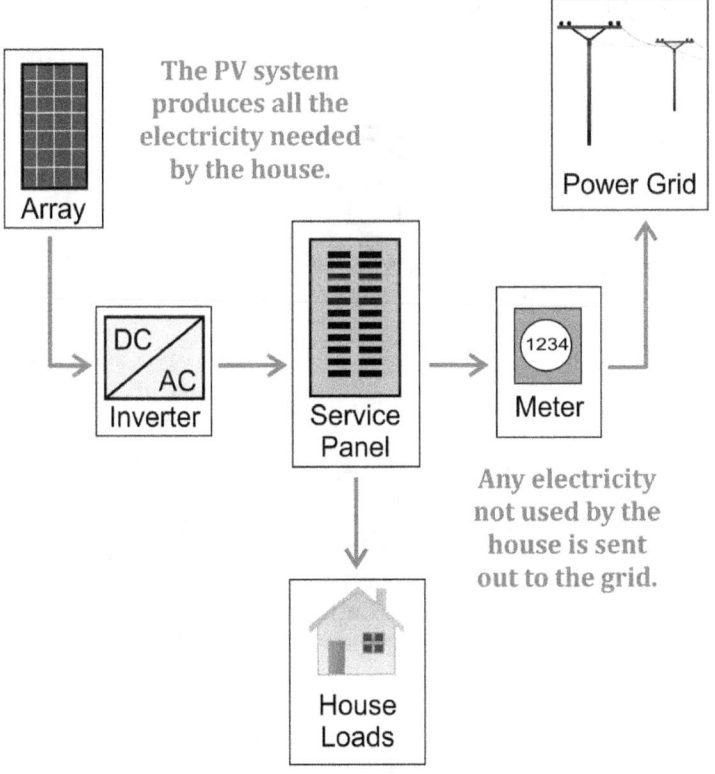

Making Solar Pay

Scenario 2: PV System Supplies Some Electricity

The house loads need more electricity than is being produced by the PV system, so the loads pull electricity from the utility's grid to make up the difference.

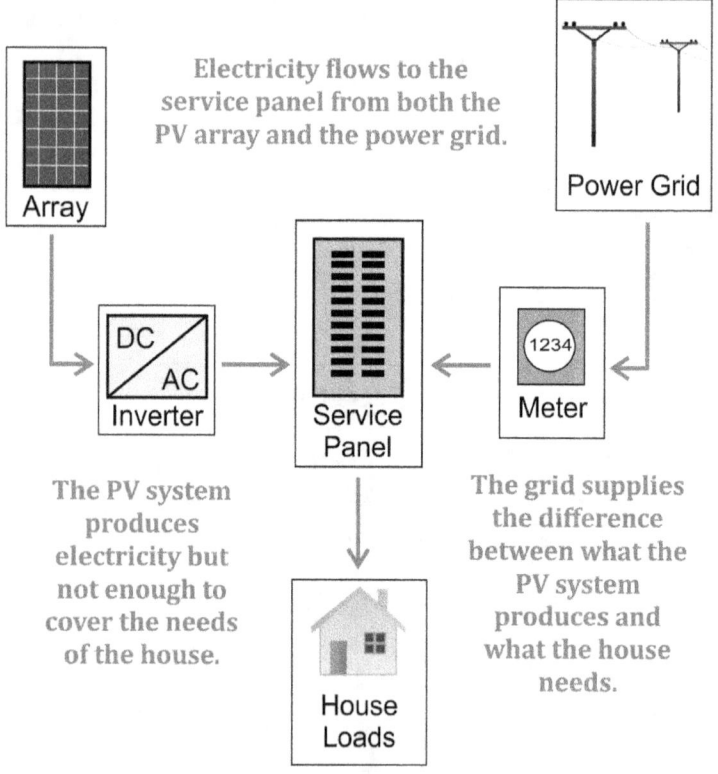

Types of PV Systems

Scenario 3: PV System Supplies No Electricity

The PV system is not generating any electricity, such as at night or during storms. The house gets all of its electrical needs from the power grid.

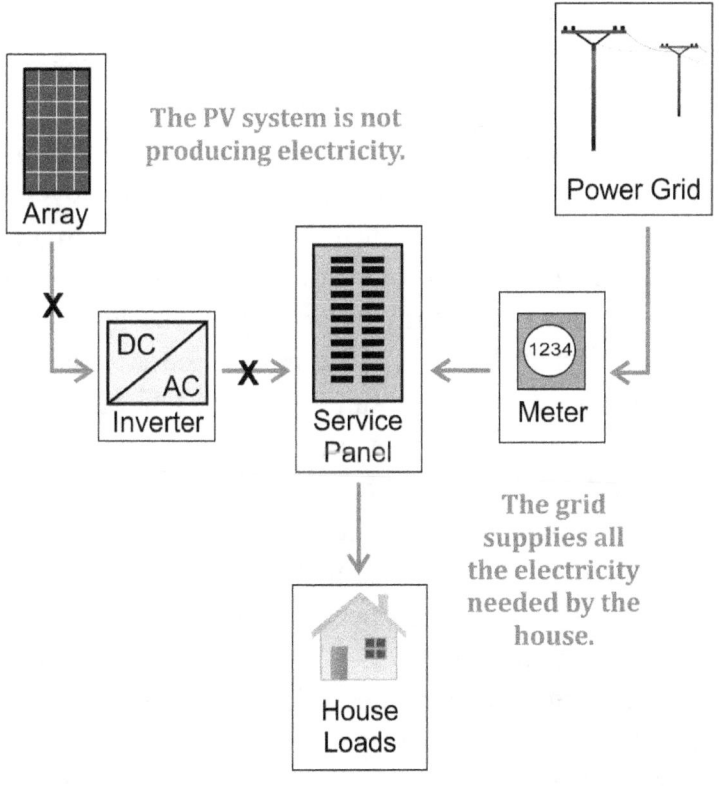

The key concept behind this design is known as **net metering**. The utility's electric meter tracks all interactions with the grid, whether you are buying (pulling power from the grid) or selling (sending it out to the grid). With net metering, the price for both transactions is the same. So if you bought 30 kWh from the grid over the course of a day, but your PV system allowed you sell 30 kWh back to the grid, then the net amount for that day would be zero.

What is interesting is that the 30 kWh produced by your PV system that day will likely be generated over a five- or six-hour period when the sun is at its optimum position in relation to your array. This time period is known as **peak sun hours**. During peak sun hours, the PV system will likely produce more electricity than is used by the house, so the surplus will be sent out to the grid as a credit. Then at night, when the system is not producing any power, you buy all the electricity that you use, but the utility automatically applies the credits that your system generated earlier in the day.

In essence what is happening is that a kilowatt-hour that you send to the utility is being stored so that you can use it at a later time for no cost. The utility grid is acting like a battery for the excess power that you create.

Another aspect of net metering is that unused credits roll over from one day to the next and will continue to increase until you use them, day by day, month by month. This enables you to balance months when electrical production is lower (generally winter, when days are shorter) with months when higher production levels allow you to bank credits. This is why solar companies want you to provide twelve months of usage data. If your goal is to have 100 percent offset of your electrical bills, your PV system needs to be designed so that at the end of the year, the cost of the energy that you purchased nearly equals the value of the credits generated.

There are several variations of a grid-direct system. The first is the simplest design whereby the modules of an array are wired together, with their electrical output being pooled and sent to one inverter. This inverter is known as a **string inverter** because the one inverter services the entire string of modules.

While this design works very well, there is an issue with it due to an oddity of PV-module production. When a group of modules are connected as one unit, the amount of electricity that can be produced by each module is limited to the lowest common production value. For example, if you have eight modules that are producing 250 watts each, then the amount of electricity being sent to the inverter is 2,000 watts.

However, if the production of one module drops to 120 watts due to creeping shade from a nearby tree, then the maximum the other seven modules can generate is also 120 watts, even if those modules are not directly impacted by the shade. So instead of 2,000 watts, you are now getting 960 watts. Taking this a step further, a malfunction in one module will shut down the entire system.

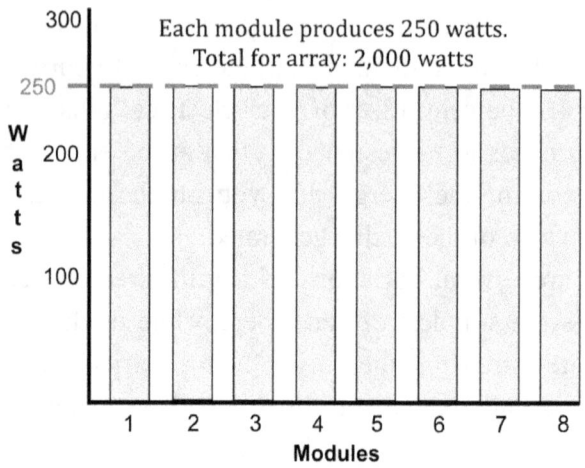

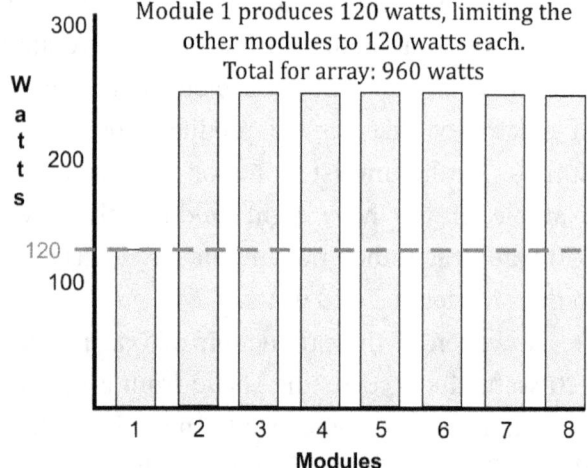

One way to fix this issue is to include **optimizers** in the design. An optimizer is a device attached to a module to regulate the power flowing from it. This reduces the

Types of PV Systems

impact of fluctuating production levels of individual modules. Using the same bar chart example where module 1 is impacted by shade, module 1's production drops to 120 watts, but the other seven modules are able to maintain their output of 250 watts. So the total production for the array would be 1,870 watts.

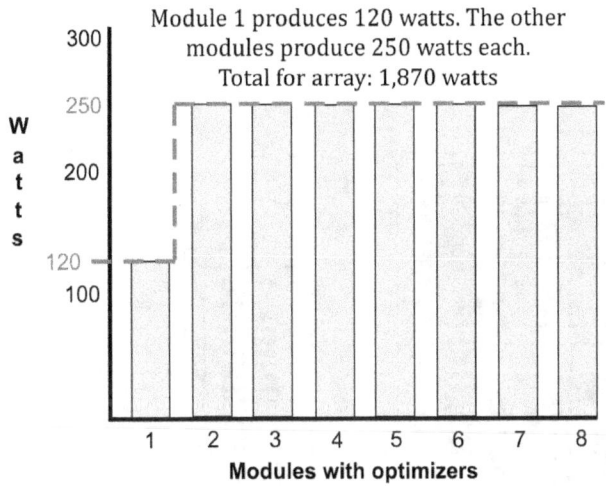

Optimizers also make it easy to monitor the performance of each module. Not only will you be able to tell immediately if production falls off, but you will be able to identify which module is causing the problem. This is discussed in more detail in the chapter on performance monitoring.

The last variation of a basic grid-direct system is one that uses **microinverters**. A microinverter is a smaller version of the standard string inverter that is attached to each module, usually underneath the module, where it is

Making Solar Pay

out of sight. In effect, each module becomes a self-contained source of power. The module produces DC electricity; the attached micro-inverter converts that to AC electricity and sends it down the line to the electrical service panel.

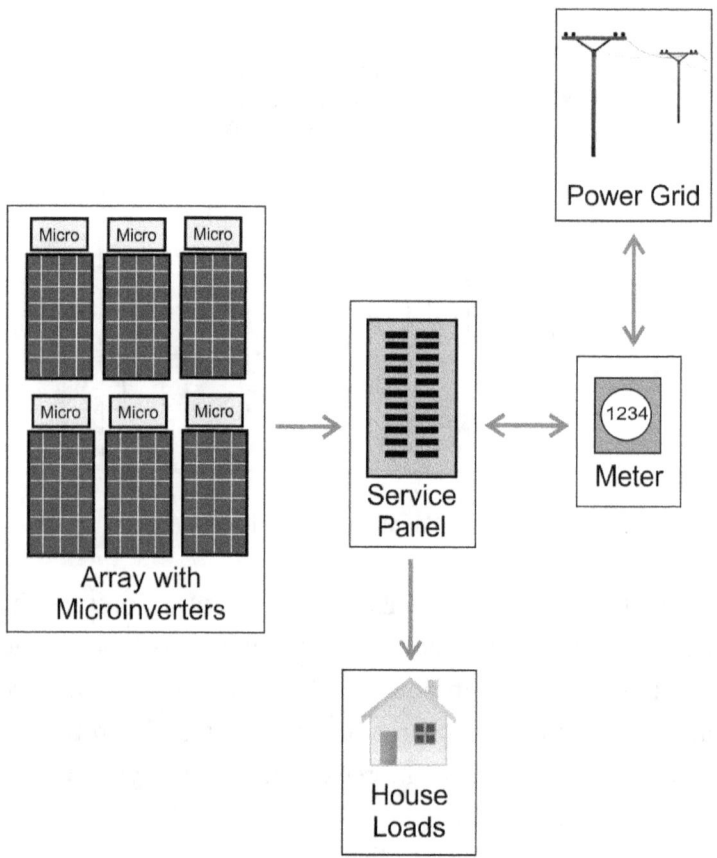

Just like optimizers, microinverters allow for in-depth performance tracking. In addition, microinverters give you the opportunity to easily increase the size of your system. With a conventional string inverter, you can add additional modules to increase capacity, but the inverter may not be large enough to handle the increased level of production. Your option at that point is to either add a second inverter or replace the existing inverter with a larger one. You don't have that issue with microinverters. Since each module is a self-contained unit of power, you can scale up production without having to worry about inverter capacity.

Before moving on to other types of PV systems, there is one misconception about grid-direct systems that I must clear up. It concerns blackouts. You would think that if the utility's grid is experiencing a blackout while the sun is shining, you should not be affected. Your PV system will operate as usual, so you will have plenty of electricity.

Uh, no.

Grid-direct systems are designed to shut down when there is a power outage. This is done for safety. Keep in mind that your PV system is a source of power not just for your house but also for the utility company. If you are sending out electricity when the grid's primary source is down, there is a very real risk that repair workers, thinking that the system is completely shut off, may be shocked or electrocuted.

If the grid has a blackout, you'll have a blackout, too. The only way to avoid that would be to have a grid-direct system that includes a battery backup. That's next.

Grid Direct with Battery Backup

The key to understanding a grid direct system with battery backup is in the name: the battery is strictly a backup source of power. This system functions as a grid-direct system when the utility grid is powered up and operating normally. The battery comes into play only when there is a blackout.

The basic design, shown on the next page, is one of the most complicated PV systems. The battery itself will likely be a bank of batteries. Several different types of batteries can be used, each type with its own life expectancy and maintenance requirements. In addition, you need a charge controller on one side of the battery to prevent overcharging and a load controller on the other side to keep the battery from being drained too low.

The design also requires two energy paths to power your home. The first is to the main service panel, which funnels electricity to most of the loads in your house. The second path is to a smaller subpanel, which has a limited number of loads. These are the ones that you identify as being crucial to stay up and running during a power outage, such as food-storage equipment, key electronic components, and certain lights. The subpanel is completely isolated from the rest of the downstream systems and devices. In other words, the subpanel has no direct connection with the main service panel or to the utility grid.

Types of PV Systems

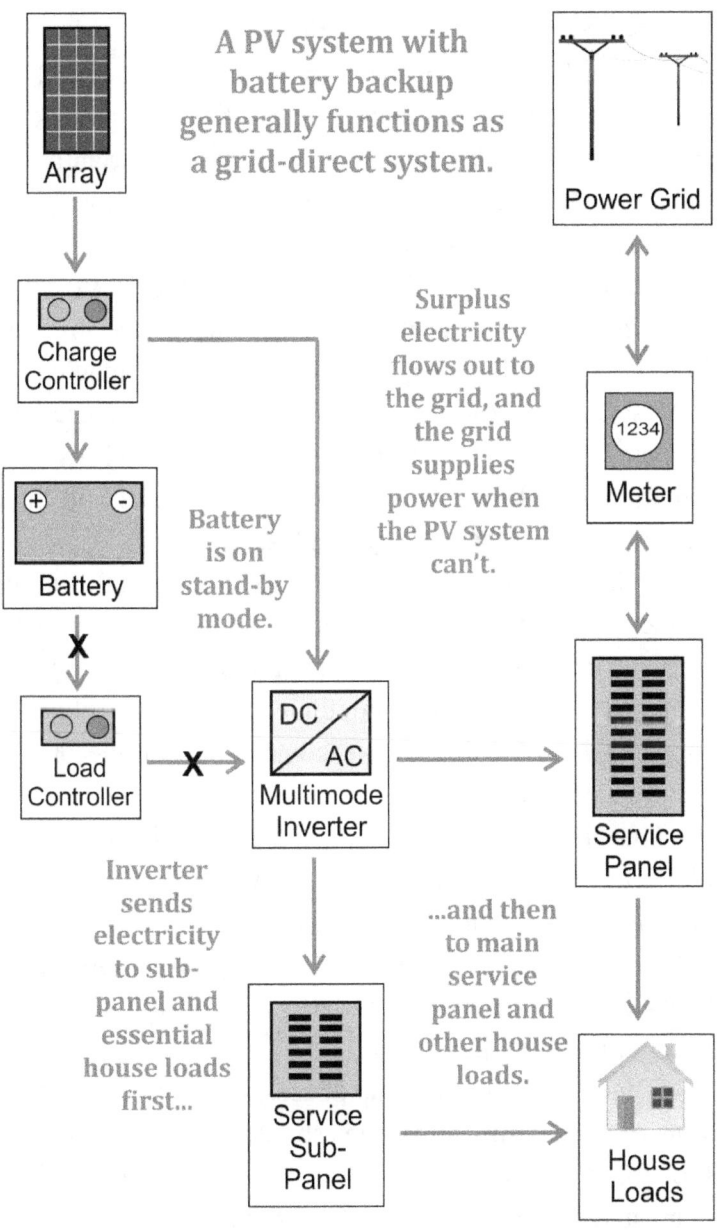

Making Solar Pay

The inverter is more complex as well. It will often include built-in features, such as a load controller and an additional component that enables the power grid to supplement charging the battery. Most importantly, it functions as a multimode inverter in that it can operate in both grid-direct and stand-alone modes.

During normal operations, the inverter is in grid-direct mode. When electrical power reaches the inverter from the array, the inverter does its DC-to-AC magic and then sends the AC power to the subpanel. Once the subpanel loads are powered, the inverter sends the remaining electricity to the main service panel and then out to the grid if there is any surplus.

In the event of a blackout, the inverter shuts down the connections to the main service panel and the grid. It then activates the battery to send the stored electricity to the subpanel loads. It will continue to power only the subpanel loads until the grid comes back online or the battery's charge is drained.

That is a key point to understand. If there is a blackout, you will not have complete power to all electrical devices in your home. The only devices that will maintain power are those that you have identified as being essential, the ones that are wired to the subpanel. Everything else will go dark.

Clearly, a battery backup adds cost and complexity to your PV system. So when should you consider it? Probably the only time it truly makes sense is if you live in an area prone to frequent power outages.

Types of PV Systems

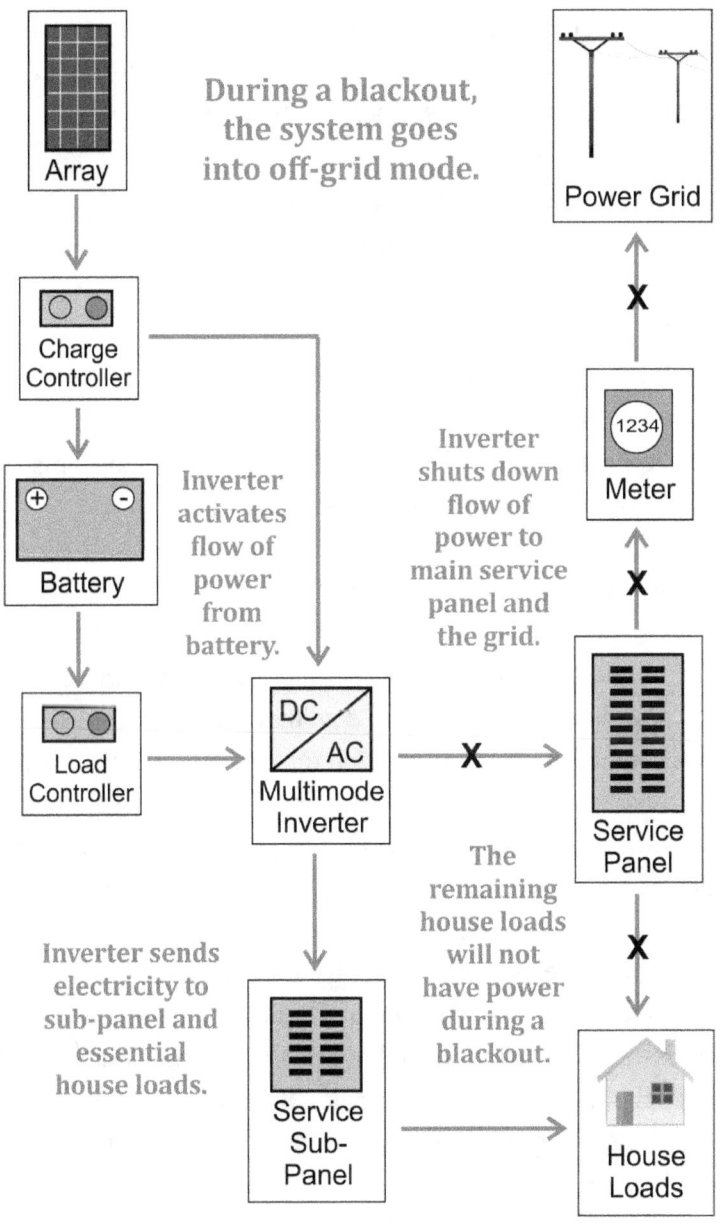

There are other types of PV systems, which often are not feasible for a typical homeowner or not readily available in many areas. These include the following:

- Stand-alone systems (off grid)
- Community solar gardens
- Zero sell
- DC direct
- Feed-in tariffs (buy all / sell all)

Appendix A has information about those systems if you wish to learn about them. See page 109.

Future Issues

A question that a lot of people ask is, Are there new developments heading our way that would cause me to wait on going solar? Solar is changing fast, but there is no real risk that your PV system will become obsolete the day after it's installed. Regardless of what changes take place, your system will still produce electricity, often guaranteed for twenty-five years.

And there is the opportunity cost of waiting. As you will see in the next chapter, some incentive programs are short term in duration or have a limit to the number of households that can participate. So an incentive that is available today may be gone tomorrow. Also, if you are paying $1,000 a year in electric bills, waiting a year means you will send another $1,000 to the utility company. You might be better off using that money to help fund a PV system.

Perhaps the main question regarding future issues should be, what are your plans? Maybe you need to account for a potential change in how much electricity you use. A home-renovation project on the horizon might lead to more square footage or provide new appliances. Or you might want to do something truly radical and install an electric car charger in your garage.

So if you have fast-approaching plans that will change your consumption pattern, now may not be the best time to go solar. From a technology standpoint, however, there is no reason to wait.

Three
Incentives

As you might expect, solar incentive programs are designed to encourage the installation of solar energy systems. Which programs you qualify for and how they work can be confusing. Hopefully, this chapter will help clarify that issue.

Renewable Energy Tax Credit

When it comes to solar incentives, the conversation always starts with the Renewable Energy Tax Credit. The tax credit is offered by the US government, and it is generous. It allows you to claim a 30 percent tax credit on the installation and qualified expenditures, which may include site preparation. The only stipulations are that the house where the solar system is installed must be within the United States and that you must use that unit as a residence, although it does not have to be your primary residence.

The program is scheduled to expire at the end of 2021, with the full 30 percent credit being available for systems placed in service by the December 31, 2019 and a gradual reduction for systems placed in service over the following two years. For a new house, "placed in service" is defined as the date you take occupancy of the residence; for an existing house, "placed in service" is the date that the solar electric system passes its final inspection and is ready for use.

The actual percentages are as follows:

- 30 percent for systems placed in service by the end of 2019
- 26 percent for systems placed in service during the 2020 calendar year
- 22 percent for systems placed in service during the 2021 calendar year

Keep in mind that this is a tax credit, not a tax deduction. A tax credit reduces your overall tax bill by the amount of the credit, as opposed to a tax deduction, which reduces the amount of income subject to tax. In other words, with a tax credit, you get to keep the whole thing; with a tax deduction, you get to keep only part of it.

As a very simplistic example, assume the installation cost of your solar energy system was $20,000 and it was placed in service on September 15, 2017. You would qualify for the 30 percent tax credit. Since 30 percent of $20,000 is $6,000, the net cost of your system drops to $14,000.

But there is a catch…of sorts. You don't get the benefits of the tax credit until you file your taxes for the year the system was placed in service. What this means is that you have to pay the full amount of the installation when the system is installed (in this case, $20,000 on September 15). You can't take advantage of the 30 percent tax credit until you file your 2017 tax returns, which would not happen until the first few months of 2018, a gap of five or six months. If you installed your system early in a calendar year, it might be fourteen or fifteen months before you get back the dough.

It might even be longer in some cases. The tax credit will offset your tax liability but only up to the amount that you owe for a given year. Any leftover tax-credit money is then carried over to the next year. If you have a $6,000 tax credit but only owe $5,000, you can apply $5,000 of the tax credit to cover your current tax liability and then carryover the remaining $1,000 to the next year. You still get the full $6,000 back; it's just that you may not get it all back in the first year.

So the timing can be difficult. You do have to pay the full amount for the installation, but in time you will get to apply a 30 percent reduction to that installation cost, and you get to keep all of it. Guaranteed.

One last note about the tax-credit program: I said earlier that the 30 percent credit applies to the cost of installation and qualified expenditures, which may include site preparations. Repairing the roof where the PV array will be mounted is generally considered a qualified expense. However, that does not mean that you can replace your entire roof and claim it as part of your solar-site preparation. Perhaps you can, but that could be a very expensive mistake if your claim ends up being denied. Do yourself a favor: check with a tax professional first.

Other Financial Incentives and Regulations

Besides the federal tax credit, you might also qualify for state and local incentives offered by government agencies or private companies. The most common types of incentives include the following:

- Performance-based incentives
- Loan programs
- Rebates
- Grants
- Property tax exemptions

The incentives and regulations vary widely from state to state, and often between cities and counties within the states. It may be best to rely on the bidding solar companies to identify the incentives you might qualify for and explain how to secure them. Since your house is in a fixed location, the companies should be telling the same story. How well a company explains the incentives to you may be another criterion to consider when deciding who gets your business.

What follows is an example of how these incentives can impact your bottom line. In this case, the electric utility offers a performance-based incentive of $0.02/kWh produced for the first ten years of production. For production purposes, I am starting with the 9,674 kWh usage number that we came up with in chapter 1. I am also assuming a one half of one percent drop in production each year, due to normal system degradation.

Incentives

	Annual kWh production	Incentive price	Total
Year 1	9,674	0.02	193.48
Year 2	9,626	0.02	192.52
Year 3	9,578	0.02	191.56
Year 4	9,530	0.02	190.60
Year 5	9,482	0.02	189.64
Year 6	9,435	0.02	188.70
Year 7	9,387	0.02	187.74
Year 8	9,340	0.02	186.80
Year 9	9,294	0.02	185.88
Year 10	9,247	0.02	184.94
		Total	**1,891.86**

The net impact of the performance-based incentive over the ten-year period is $1,891.86. When you factor that in with the example we used for the Renewable Energy Tax Credit, you get the following:

Cost of installation	20,000
less 30% tax credit	6,000
less performance-based incentive	1,892
Net cost	**12,108**

So from an initial installation cost of $20,000, the net cost of the PV system ends up being $12,108, a savings of just over 39 percent.

This is looking pretty good.

Four
Buying Options

Before diving into analyzing proposals, I want to cover your options for acquiring a solar PV system. You have three:

1. Buying using cash
2. Financing with a loan
3. Leasing

Let's take a look at all three options and consider the pros and cons of each.

Buying Using Cash

As with most purchases, cash is king. It is the simplest transaction to implement and the cheapest option in the long run. Some companies even offer additional discounts on the final price if you pay in cash. That's just one way of thanking you for making their lives easier too.

Buying with cash also leads to the highest net impact on your electric bill. All of the savings generated by the PV system go straight into your pocket. Over the life of the system, that adds up to a large sum and leads to a complete payback on your investment in the shortest amount of time.

Owning the system outright can add substantial value to your house, as well. Real estate appraisers use several methods to assess the value that a PV system adds to a home. We will get into those in the chapter on financial analysis. For now, suffice it to say that using cash will have a significant impact on your break-even point.

The downside of using cash is twofold. First, it entails a large outlay of money. As discussed in the chapter on incentives, you will get a 30 percent tax credit, but you won't see it until you file a tax return for the year in which the PV system was installed. That time period between paying the cash and getting the tax credit generally ranges from several months to just over a year.

Second, there is the issue of what else you could have done with that money, the elusive lost opportunity. Many compare installing solar with remodeling your kitchen or adding a new bathroom. It is important to note that a solar PV system is not just another home improvement project. Unlike a kitchen or bathroom, a PV system generates revenue. You will be selling energy to the utility company every day that the sun shines. So more accurately, a PV system is an investment in electrical energy. And you should expect to get back every dollar that you put into it.

If you truly want to look at solar energy's opportunity cost, you should compare it with other revenue-producing vehicles, such as stocks, bonds, and similar investments. The key issue is whether those other investments offer higher rates of return. Comparing rates of return is a complicated business, requiring several variables such as target portfolio makeup, annual percentage yield, discount rates, reinvestment rates, and risk tolerance, to name a few. A financial advisor is the best person to discuss what you might be giving up by investing in solar. However you look at it, though, you will likely find that a solar investment will give you a safe, conservative return on your money.

Financing with a Loan

One way to avoid the large upfront cost when purchasing a PV system is to use a loan, such as a bank loan, a mortgage refinance, a home equity line of credit, or some other instrument. Of course, the trade-off is that you will have monthly loan payments.

The key to making this work is that the amount that you save each month on your electric bill needs to be higher than the amount that you pay to service the loan. If you can arrange that, the system will pay for itself in time. The length of time it takes to get to breakeven will be much longer than the cash option, but you will get there eventually. Assuming that you have a fixed interest payment, the net amount that you save per month will go up as the inevitable rise in utility rates will increase your savings.

Another thing to consider is that you will have all the benefits of ownership while you are paying off that loan. You are entitled to the 30 percent Renewable Energy Tax

Credit and all other incentives tied to the PV system. You will also benefit from any value-add impact on your home's value.

Finally, borrowing money to finance a PV system enables you to use the cash that you didn't spend on some other investment. If the overall return on your investment portfolio is higher than the return on the PV investment, then financing your PV system makes a lot a sense.

Leasing

A third way to acquire a PV system is through a lease. With a lease, you make a small payment upfront to initiate the program, the leasing company installs the PV system on your property, and then you will have monthly lease payments. In exchange for that, you can use the electricity the system produces to offset the electricity that you would normally buy from the utility. And just like with a system that you bought with cash or a loan, any energy that you don't use is sent out to the grid for a full credit, as prescribed in your local net-metering regulations.

The key concept behind a lease is that the leasing company retains ownership of the equipment. There is good news and bad news with that. The good news is that the leasing company is responsible for keeping your system up and running, so you have no maintenance costs. However, a PV system has no moving parts, so the maintenance associated with it is very low.

The bad news is that you will not qualify for any of the financial incentives associated with a PV system. By retaining ownership, the leasing company is entitled to the federal tax credit and all other discounts and rebates offered by the state and local governments and utilities.

Worse news is that a leased PV system will not add significant value to your home. From a real estate appraiser's perspective, a leased PV system is the leasing company's asset, not yours. Thus, the value of the system cannot be included in a formal appraisal.

There are several questions you should ask before signing a lease:

- **Can you terminate the lease early?** The lease is a financial obligation for a set period of time. It is important to know whether you have the option to cancel the lease early. If you do have that option, you need to find out what the cost and penalties would be if you did that.

- **Are your lease payments fixed, or will they rise over time?** Some lease agreements allow lease payments to track rising utility costs. If the cost of electricity goes up, so will your monthly expense. You will likely still be paying less than if you were buying all your electricity from the utility, but you can't escape the fact that the overall cost of the lease just went up.

- **What happens at the end of the lease?** Usually you can buy the PV system at the end of the lease. If the lease is for twenty years, the PV system will likely continue to produce electrical savings for an additional five to ten years, possibly more. So there may be value you can take advantage of. The other alternative is to have the PV equipment removed. Check to see if there would be a cost for that removal. Also, you should expect to need roof repairs after the system is gone.

- **Who covers insurance for the leased equipment?** You don't own the system, but it is installed on your property. Some lease agreements stipulate that you must add the system to your homeowner's insurance. That could result in an increase in premiums you pay for insurance, which would lower the financial benefit that you would be getting from the lease.

- **Is the lease transferrable if you sell the house?** Many leases can be transferred to the new buyer. However, you may still be obligated if the buyer doesn't want to play or doesn't get approval from the leasing company. In the event of either one of those occurrences, you may have to terminate the lease and have the system removed (which takes us back to the first question).

A variation of a lease is the power purchase agreement (PPA). With a PPA, you don't lease the equipment; you just agree to purchase all of the power generated by the equipment at a set price. Like a lease, the PV system is installed on your property, so the same issues and questions apply. The primary advantage of a PPA is that the leasing company has an incentive to make sure that the system is operating at peak efficiency. You have agreed to buy all the power produced by the system. The more it makes, the more you buy. Whatever you don't use, you can send out to the grid for a credit via net metering. But just as with a lease, the company behind the PPA owns the PV system and gets all the benefits that go with ownership.

Comparing the Options

The table below shows the impact of several issues in relation to using cash, taking out a loan, and leasing a PV system.

Comparison of Cash, Loan and Lease Options

	Cash	**Loan**	**Lease**
Finance charges	No	Yes	Yes
Renewable Energy Tax Credit	Yes	Yes	No

	Cash	**Loan**	**Lease**
Other incentives	Yes	Yes	Possibly, but most will go to the leasing company since they own the system.
Upfront costs	High. You have to pay the entire cost of the system up front.	None. You borrow the money instead of paying cash.	Little to none. There may be a small payment to initiate the lease.
Net savings on electric bill	The entire net-metered amount produced by your system.	The savings produced by your system less the loan payment.	The savings produced by your system less the lease payment.
Ownership of system	Yes	Yes, once the loan is paid off. Until then, however, you get all benefits of ownership.	No, although you may have the option to purchase at the end of the lease.

Buying Options

	Cash	**Loan**	**Lease**
Maintenance costs	Yes, minimal upkeep expenses. And you may have to replace the inverter around Year 15.	Yes, same as cash option.	None, although there may be some ongoing expenses, such as insurance. But you probably won't have to pay to replace the inverter.
Adding value to your home	Yes. You own the system. It can be included in an appraisal of your home.	Yes, same as cash option.	No. You do not own the system, so it cannot be considered as part of the value of your home.
Deductibility of interest payments	No interest payments means there's nothing to deduct.	Interest payments are tax deductible if you are financing solar as part of a mortgage.	There is no interest in a lease payment so there is no tax deduction.

	Cash	**Loan**	**Lease**
Payback time	Quickest payback; the only cost you have to overcome is the actual cost of the system.	Longer payback due to finance charges increasing the overall cost of the system.	There is no payback because you don't own the system. Lease payments are strictly expense.
Transferable	Yes. You own the system, so you can sell it as part of your house.	Yes. You can sell the system as part of your house, but you will have to pay off the loan.	Generally, leases are transferable. Be sure to verify this with your leasing company.
Escalating costs	None	None, provided you have a fixed-rate loan.	Generally, none, but some lease payments can go up to track rising utility costs.

Five
Proposals

Proposals from companies bidding for your business can have varying degrees of detail, ranging from somewhat skimpy to very complex. At a minimum, though, they all should have certain data points in common. This chapter will help you organize those data points so that you can do a meaningful comparison.

Gathering the Essential Information

Most proposals will have the following categories of information:
- System
- Cost
- Environmental
- Financial

I will get to system and cost in a moment. The environmental section provides the benefits of going solar, showing how a PV system reduces your carbon footprint (the environmental impact of fossil fuels being used to generate your energy). Generally, this information is presented as comparisons, such as trees saved, homes powered, gas not used, and such. While those details are interesting, they should not be factors in

Making Solar Pay

your decision. You should not select one company over another simply because a proposal states it will save ten more trees over a twenty-five-year period. You will have a positive impact on the environment regardless of which company you choose.

A proposal's financial section is designed to present the economic case for going solar. That is key information and should be used in your decision. But it is premature to consider it when you are in the "comparing proposals" stage. It is better to determine which proposal is best and then work through that proposal's financial implications to see if solar makes sense for you. The next chapter shows you how to do that.

To analyze competing proposals, you should focus on the system and cost data. For this analysis, we will use a spreadsheet that you can find at the Mountain Edge Publishing website (www.mepub.com). You enter data for the highlighted variable inputs.

Proposal Comparison

Goal: 1   Number of companies bidding: 1

First Cut: Analyze Cost

	Company 1	Company 2	Company 3	Company 4	Company 5
System					
Proposed size (kW)	1.00	1.00	1.00	1.00	1.00
Average peak sun hours	0.00	0.00	0.00	0.00	0.00
Estimated production	0	0	0	0	0
% offset	0.0%	0.0%	0.0%	0.0%	0.0%
Cost					
Cost	0	0	0	0	0
Cost per watt	0.00	0.00	0.00	0.00	0.00

Proposals

The first two are numbers you provide; the other three should be in each company's proposal:

- **Goal**: the number that you gave the companies to build their proposals, derived from square-footage estimates, utility usage reports, or previous electric bills.

- **Number of companies**: How many bids will you compare?

- **Proposed size (kW)**: the size of the system a company is proposing to build.

- **Estimated production**: the production number that a company proposes to deliver.

- **Cost**: your total out-of-pocket expense to get the system up and running. It might be shown in the proposal as Total System Cost, Gross Purchase Price, Amount Payable, or something like that. It's the price before accounting for incentives that will take several months or years to fully realize, such as tax credits, anticipated savings, and performance incentives.

Where to Find the Data in a Proposal: Example 1

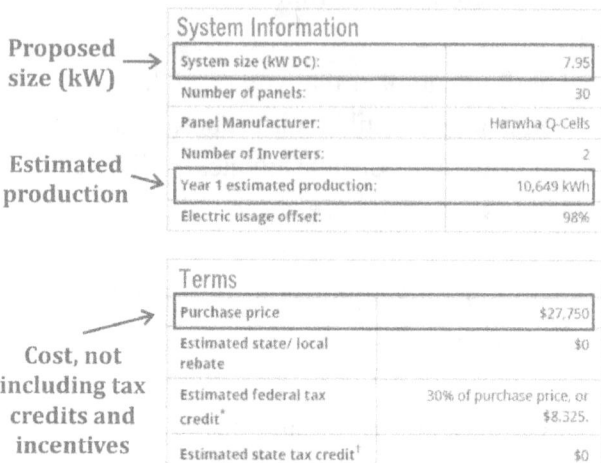

Where to Find the Data in a Proposal: Example 2

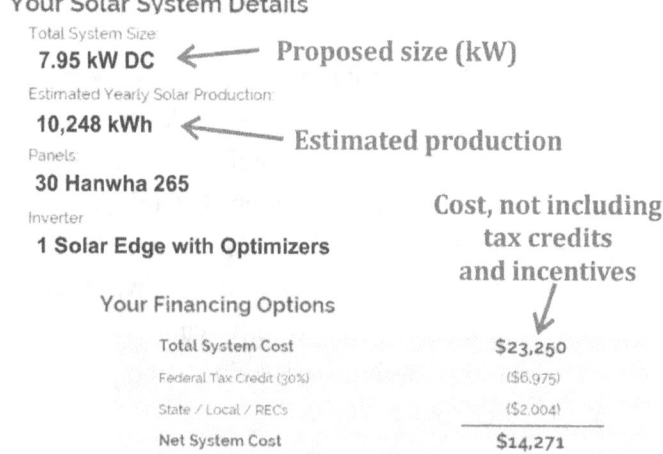

The non-highlighted fields are the results of calculations using the values that you entered:

- **Average peak sun hours**: As mentioned in chapter 2, average peak sun hours is the potential hours of maximum sunlight per day. It helps determine production estimates: proposed size * average peak sun hours * 365 days = annual production estimate.

- **% offset**: shows how well the proposed PV system will meet your stated goal. It's a simple calculation: production estimate / goal = % offset.

- **Cost per watt**: another simple calculation: cost / proposed size = cost per watt. The proposed size is in kilowatts (kW), so the first step is converting kilowatts to watts (1 kW = 1,000 watts).

Before beginning any analysis, you need assurance that this exercise will be useful. One way to do that is to look at average peak sun hours. Those numbers should be about the same for each proposal because your house is a fixed entity in a known location. If the number used by one company is significantly higher, that company could show a similar production estimate as the other proposals but with a smaller system and subsequently a lower price tag.

Proposal Comparison

Goal: 9,674

Higher "average peak sun hours" can lead to comparable production with a smaller system.

First Cut: Analyze Cost

	Company 1	Company 2	Company 3	Company 4	Company 5
System					
Proposed size (kW)	6.15	6.90	6.95	6.80	6.85
Average peak sun hours	4.33	3.83	3.82	3.85	3.82
Estimated production	9,725	9,638	9,699	9,545	9,555
% offset	100.5%	99.6%	100.3%	98.7%	98.8%
Cost					
Cost	18,742	19,941	23,226	23,868	20,148
Cost per watt	3.05	2.89	3.34	3.51	2.94

When the companies have roughly the same number for average peak sun hours, you can fairly compare proposed size and estimated production, which in turn drives a fair cost comparison. At a glance, you can see that the numbers in the example are close, so everything looks good.

Proposal Comparison

Goal: 9,674

Number of companies bidding: 5

First Cut: Analyze Cost

	Company 1	Company 2	Company 3	Company 4	Company 5
System					
Proposed size (kW)	6.75	6.90	6.95	6.80	6.85
Average peak sun hours	3.95	3.83	3.82	3.85	3.82
Estimated production	9,725	9,638	9,699	9,545	9,555
% offset	100.5%	99.6%	100.3%	98.7%	98.8%
Cost					
Cost	20,567	19,941	23,226	23,868	20,148
Cost per watt	3.05	2.89	3.34	3.51	2.94

First Cut: Analyze Cost

The first step of the analysis is to compare the cost per watt. That one figure should encompass everything. For that price, the company is saying that it can install

Proposals

and activate a PV system that should meet the production estimates and impact your annual electrical usage by the percentage offset.

The nice thing about having competing bids is you don't have to overpay. You can eliminate any company whose cost-per-watt figure is too high relative to the other proposals. In the example below, you can see that two companies are significantly higher than the other three.

Proposal Comparison

Goal	9,674		Number of companies bidding	5	
First Cut: Analyze Cost					
	Company 1	Company 2	Company 3	Company 4	Company 5
System					
Proposed size (kW)	6.75	6.90	6.95	6.80	6.85
Average peak sun hours	3.95	3.83	3.82	3.85	3.82
Estimated production	9,725	9,638	9,699	9,545	9,555
% offset	100.5%	99.6%	100.3%	98.7%	98.8%
Cost					
Cost	20,567	19,941	23,226	23,868	20,148
Cost per watt	3.05	2.89	3.34	3.51	2.94

The differences are high enough that those two companies can be eliminated.

Proposal Comparison

Goal	9,674		Number of companies bidding	5	
First Cut: Analyze Cost					
	Company 1	Company 2	~~Company 3~~	~~Company 4~~	Company 5
System					
Proposed size (kW)	6.75	6.90	~~6.95~~	~~6.80~~	6.85
Average peak sun hours	3.95	3.83	~~3.82~~	~~3.85~~	3.82
Estimated production	9,725	9,638	~~9,699~~	~~9,545~~	9,555
% offset	100.5%	99.6%	~~100.3%~~	~~98.7%~~	98.8%
Cost					
Cost	20,567	19,941	~~23,226~~	~~23,868~~	20,148
Cost per watt	3.05	2.89	~~3.34~~	~~3.51~~	2.94

Making Solar Pay

Second Cut: Level Production Estimates

After making the first cut, you can further analyze the remaining candidates by leveling the production estimates. We've already talked about how the solar companies reached their production estimates by making assumptions on the average peak sun hours. The question is, how would those estimates look if they all used the same sun-hours number?

The spreadsheet makes that calculation for you. It takes the average number for all the peak sun hours, shown at the right of the hours of individual companies, and inserts that figure in the corresponding field for each company in the bottom section. It then recalculates estimated production and % offset.

Proposal Comparison

Goal	9,674		Number of companies bidding	5		

First Cut: Analyze Cost

	Company 1	Company 2	~~Company 3~~	~~Company 4~~	Company 5	
System						
Proposed size (kW)	6.75	6.90	~~6.95~~	~~6.80~~	6.85	
Average peak sun hours	3.95	3.83	~~3.82~~	~~3.85~~	3.82	3.85
Estimated production	9,725	9,638	~~9,699~~	~~9,545~~	9,555	
% offset	100.5%	99.6%	~~100.3%~~	~~98.7%~~	98.8%	
Cost						
Cost	20,567	19,941	~~23,226~~	~~23,868~~	20,148	
Cost per watt	3.05	2.89	~~3.34~~	~~3.51~~	2.94	

Second Cut: Level Production Estimates

	Company 1	Company 2	~~Company 3~~	~~Company 4~~	Company 5
System					
Proposed size (kW)	6.75	6.90	~~6.95~~	~~6.80~~	6.85
Average peak sun hours	3.85	3.85	~~3.85~~	~~3.85~~	3.85
Estimated production	9,493	9,704	~~9,774~~	~~9,563~~	9,633
% offset	98.1%	100.3%	~~101.0%~~	~~98.9%~~	99.6%
Cost					
Cost	20,567	19,941	~~23,226~~	~~23,868~~	20,148
Cost per watt	3.05	2.89	~~3.34~~	~~3.51~~	2.94

Proposals

Now with all things being equal, you can see how close each company comes to meeting your goal and at what cost. In the example below, Company 2 and Company 5 have similar production estimates, % offsets, and costs per watt. Company 1 is offering a slightly lower production number at a higher cost. Company 1 can be eliminated.

Second Cut: Level Production Estimates

	Company 1	Company 2	~~Company 3~~	~~Company 4~~	Company 5
System					
Proposed size (kW)	6.75	6.90	~~6.95~~	~~6.80~~	6.85
Average peak sun hours	3.85	3.85	~~3.85~~	~~3.85~~	3.85
Estimated production	9,493	9,704	~~9,774~~	~~9,563~~	9,633
% offset	98.1%	100.3%	~~101.0%~~	~~98.9%~~	99.6%
Cost					
Cost	20,567	19,941	~~23,226~~	~~23,868~~	20,148
Cost per watt	3.05	2.89	~~3.34~~	~~3.51~~	2.94

You are now down to two viable options.

Second Cut: Level Production Estimates

	~~Company 1~~	Company 2	~~Company 3~~	~~Company 4~~	Company 5
System					
Proposed size (kW)	~~6.75~~	6.90	~~6.95~~	~~6.80~~	6.85
Average peak sun hours	~~3.85~~	3.85	~~3.85~~	~~3.85~~	3.85
Estimated production	~~9,493~~	9,704	~~9,774~~	~~9,563~~	9,633
% offset	~~98.1%~~	100.3%	~~101.0%~~	~~98.9%~~	99.6%
Cost					
Cost	~~20,567~~	19,941	~~23,226~~	~~23,868~~	20,148
Cost per watt	~~3.05~~	2.89	~~3.34~~	~~3.51~~	2.94

Other Things to Consider

Besides numbers, there are a few other factors that can help you make a final selection.

- **Warranties**: There are a few questions to ask about warranties. What are the warranty time frames for the various components? Who is actually guaranteeing the system, the installation company or the manufacturer? Is the warranty transferrable?

- **Assumptions**: We've talked about the ripple effect that assumptions can have. You want to make sure that a company's proposal is better because it really is better, not just because the underlying assumptions are skewing the data in one direction. However, don't spend too much time on the financial projections; as I said earlier, chapter 6 and the appendices show you how to use spreadsheets to easily make those projections yourself.

- **Add-ons**: Check to see if one company is including things that others are not. These might be enhancements such as production-monitoring devices and other latest/greatest technology marvels. Or they could be useful low-tech items such as animal guards to keep out nesting and gnawing critters.

- **References**: Talking with past customers is probably not the most productive exercise. No company is going to give you the name and number of someone who was dissatisfied. But you might get a feel for how close a proposed production estimate was to actual production numbers. If the estimate was way off (higher or lower), then you might want to factor that in.

- **Schedules**: The last piece you may want to consider is schedule, especially when you are heading into the last few months of the year. Remember, the key date for the federal tax credit is when the PV system has passed its final inspection and is ready for use. Should the system be installed in late December but the final inspection doesn't take place until January, you will have to wait about fifteen months to get the tax credit. If another company can complete the work and get the final inspection before the end of the year, you would receive the tax credit in about three months, a full year earlier. In that case, you might choose a company with a slightly higher bid, provided you can reap the tax benefits much sooner.

After you've narrowed your choice to one or two companies, it's time to make sure that the proposed PV system makes sense financially. Please go to the next chapter.

Six
Financial Analysis

Solar is an investment in electrical energy. As such, it has to make financial sense. This chapter will help you determine that, whether you accept the financial conclusions in a solar company's proposal or analyze the data yourself.

The chapter covers two basic scenarios: using cash and borrowing the money (either by taking out a dedicated solar loan or by rolling the cost of solar into a mortgage). Both scenarios imply owning the PV system. When evaluating either scenario, you should consider two factors:

- **Cash flow**: tracking solar's revenue and expenses

- **Value add**: identifying the potential impact on your home's appraised value

There is nothing wrong with relying on the financial data provided by a solar company. One problem, though, is if you have five different proposals, you may face five vastly different presentations of that data. Even with the most lucid presentation, it can be hard to follow how they made their calculations. And you always have to take into account the underlying assumptions, which can have a huge ripple effect on projected estimates.

Perhaps a better way to go is to conduct your own analysis, where you control the assumptions and numbers that drive the financial calculations. This book has three appendices that show how to do that, using downloadable spreadsheets:

- Appendix B: Cash Option
- Appendix C: Solar Loan Option (utilizing a dedicated solar loan)
- Appendix D: Mortgage Option (adding the cost of solar to a mortgage)

It's not as daunting as it may seem. You plug a few inputs into a spreadsheet, and the spreadsheet does the work for you. Then all you have to do is look at two sets of numbers to know whether going solar makes financial sense.

One last note before we dive into this: the chapter does not include a leasing scenario. You don't own a leased PV system. It may lower your electric bills, but the real value of the system will go to the leasing company, including all tax credits and financial incentives linked to ownership. A leased system will have virtually no impact on the appraised value of your home, which, you will see, is a key aspect of solar's viability.

As mentioned above, any financial analysis should focus on two key areas: cash flow and value add. Let's start with cash flow.

Financial Analysis

Cash Flow

Cash flow follows the money in and out. How you view cash flow depends on whether you purchase your PV system with cash or by borrowing the money.

Cash Option: Cash Flow

When you buy with cash, your biggest outlay by a wide margin is your upfront cost: the cost of getting the PV system up and running. There will likely be some maintenance expense every now and then, but a PV system has no moving parts, so very little maintenance is required.

For the most part, it will be money coming in. In year 1, you have the 30 percent tax credit, and then you may have some additional revenue from performance-based incentives and other programs. Most importantly, there are the annual savings on your electric bill, which is likely your main financial reason for going solar.

The money that you make with a PV system is money that you did not spend, often referred to as an avoided cost. It is determined by multiplying the amount of electricity that you produce in a year by the current rate charged by the utility company. Those savings should increase over time as the utility's electrical rate will likely rise faster than your PV system's production rate will drop.

Making Solar Pay

Your savings will go up over time.

Year	Production	Utility Elec Rate	System Cost	Federal Tax Credit	PBIs & Other Incentives	Annual Elec Savings	Net Cash Flow
Year 0			19,941				-19,941
Year 1	9,696	0.1226		5,982	194	1,189	7,315
Year 2	9,648	0.1269				1,224	1,365
Year 3	9,600	0.1313				1,261	1,398
Year 4	9,552	0.1359				1,298	1,432
Year 5	9,504	0.1407				1,337	1,467
Year 6	9,456	0.1456				1,377	1,502
Year 7	9,409	0.1507				1,418	1,539
Year 8	9,362	0.1560				1,460	1,577
Year 9	9,315	0.1614				1,504	1,616
Year 10	9,269	0.1671				1,549	1,657
Year 11	9,222	0.1729				1,595	1,514
Year 12	9,176	0.1790				1,642	1,557
Year 13	9,130	0.1853				1,691	1,602
Year 14	9,085	0.1917				1,742	1,648
Year 15	9,039	0.1985				1,794	-2,100
Year 16	8,994	0.2054				1,847	1,744
Year 17	8,949	0.2126				1,902	1,794
Year 18	8,904	0.2200				1,959	1,845
Year 19	8,860	0.2277				2,018	1,898
Year 20	8,815	0.2357				2,078	1,952
Year 21	8,771	0.2439				2,140	2,007
Year 22	8,727	0.2525				2,204	2,065
Year 23	8,684	0.2613				2,269	2,123
Year 24	8,640	0.2705				2,337	2,184
Year 25	8,597	0.2799				2,407	2,246
					1,896	43,242	

Production decreases slightly each year.

But that is offset by increases in the rate charged by the utility...

...which leads to an increase in the amount you save each year.

And that brings us to Running Total, which gives an accrued perspective of the cash flow numbers. The Running Total for year 0 (the year the PV system is installed) is the net investment, the sum of your upfront costs. Then for each year, Running Total tracks your progress in chipping away at that net investment figure. When the number goes from negative to positive, you've recouped 100 percent of the money that you invested. In the example on the next page, the Running Total goes positive in year 10, taking roughly 9.4 years.

Financial Analysis

	Production	Utility Elec Rate	System Cost	Federal Tax Credit	PBIs & Other Incentives	Annual Elec Savings	Net Cash Flow	Running Total
Year 0			-19,941				-19,941	-19,941
Year 1	9,696	0.1226					7,315	-12,626
Year 2	9,648	0.1269					1,365	-11,261
Year 3	9,600	0.1313					1,398	-9,863
Year 4	9,552	0.1359					1,432	-8,432
Year 5	9,504	0.1407					1,467	-6,965
Year 6	9,456	0.1456			189	1,377	1,502	-5,463
Year 7	9,409	0.1507			188	1,418	1,539	-3,923
Year 8	9,362	0.1560			187	1,460	1,577	-2,346
Year 9	9,315	0.1614			186	1,504	1,616	-729
Year 10	9,269	0.1671			185	1,549	1,657	927
Year 11	9,222	0.1729					1,514	2,441
Year 12	9,176	0.1790					1,557	3,998
Year 13	9,130	0.1853					1,602	5,600
Year 14	9,085	0.1917					1,648	7,248
Year 15	9,039	0.1985					-2,100	5,148
Year 16	8,994	0.2054				1,847	1,744	6,892
Year 17	8,949	0.2126				1,902	1,794	8,685
Year 18	8,904	0.2200				1,959	1,845	10,530
Year 19	8,860	0.2277				2,018	1,898	12,428
Year 20	8,815	0.2357				2,078	1,952	14,380
Year 21	8,771	0.2439				2,140	2,007	16,387
Year 22	8,727	0.2525				2,204	2,065	18,452
Year 23	8,684	0.2613				2,269	2,123	20,575
Year 24	8,640	0.2705				2,337	2,184	22,759
Year 25	8,597	0.2799				2,407	2,246	25,005
					1,896	43,242		

Running Total adds in Net Cash Flow each year to show where you are financially.

When Running Total goes positive, you have recouped your entire solar investment.

That may seem like a long time, but consider this: 100% investment / 9.4 years = 10.6% per year. So you are getting back 10.6 percent of your money each year. Technically, that is not your return on investment. ROI is a complicated calculation because of the variables that you have to account for, such as annual percentage yield, discount rates, reinvestment rates, and risk tolerance. The calculation shown here is simple, and while not a real ROI, it is a meaningful number.

There is one last point to make about solar payback. Since the money comes to you as savings instead of revenue, it is not taxable. If you ever compared solar's true ROI with the ROI of other investments (which you could do with the help of a financial advisor), you would need to consider the overall tax implications. A PV investment often ends up being very competitive, due in part to not creating a tax liability.

Loan Option: Cash Flow

With a loan, cash flow's main concern is that you are **net positive**, meaning that the amount you save is greater than the amount you pay to service the loan. Just like with the cash option, the amount you save each year should go up due to electric rates rising faster than drops in production.

In the spreadsheet example, you are net positive in year 1 and all years after that, except year 15, when you have to replace the inverter. Once the loan is paid off, you will be net positive to the tune of over $2,000 per year, assuming a very conservative escalation rate of the utility's charge per kilowatt-hour.

The point to take from this is that as long as you are net positive within the first couple years and then stay net positive after that, you will recoup your out-of-pocket expenses and pay off the loan, all the while spending less each year than you would if you had not gone solar.

If you rolled the cost of a PV system into a mortgage, you have the added benefit of all interest paid being tax deductible. That will have only a slight impact since the solar part of your mortgage payment is small. But even that will help your bottom line.

Financial Analysis

	Finance Charges	Main-tenance	Replace Inverter	Federal Tax Credit	Bridge Payoff	PBIs & Other Incentives	Annual Elec Savings	Net Cash Flow	Running Total
Year 0								0	0
Year 1	-1,190	-50		5,982	-5,982	194	1,189	143	142
Year 2	-1,190	-52					1,224	175	317
Year 3	-1,190	-55					1,261	208	525
Year 4	-1,190	-58					1,298	242	767
Year 5	-1,190	-61					1,337	276	1,043
Year 6	-1,190	-64					1,377	312	1,355
Year 7	-1,190	-67					1,418	349	1,705
Year 8	-1,190	-70					1,460	387	2,092
Year 9	-1,190	-74					1,504	426	2,519
Year 10	-1,190	-77					1,549	467	2,985
Year 11	-1,190	-81					1,595	324	3,309
Year 12	-1,190	-85					1,642	367	3,676
Year 13	-1,190	-90					1,691	412	4,088
Year 14	-1,190	-94					1,742	458	4,546
Year 15	-1,190	-99	-3,795				1,794	-3,290	1,256
Year 16	-1,190	-104					1,847	554	1,809
Year 17	-1,190	-109					1,902	604	2,413
Year 18	-1,190	-114					1,959	655	3,068
Year 19	-1,190	-120					2,018	708	3,775
Year 20	-1,190	-126					2,078	762	4,537
Year 21		-132					2,140	2,007	6,545
Year 22		-139					2,204	2,065	8,609
Year 23		-146					2,269	2,123	10,733
Year 24		-153					2,337	2,184	12,917
Year 25		-161					2,407	2,246	15,163
						1,896	43,242		

You are net positive in Year 1 and remain net positive each year except for when you replace the inverter.

The amount you save increases substantially once the loan is paid off.

Value Add: Impact of Solar on Appraised Value

The first thing to understand about real estate appraisals is that an appraiser must account for existing market conditions. In a hot market, the estimated value of a PV system probably won't matter. If multiple buyers want a house, somebody is going to pay a premium price, regardless of whether the house has solar. In a down market, the PV system may add value, but it probably won't save the house from being sold at a reduced price. The real estate market is very fluid, and the market always rules.

Making Solar Pay

A solar PV system is an asset that generates revenue, so there is no doubt that it will add value to a house. Typically there are three methods that appraisers use to determine that value: sales comparison, cost, and income. An appraiser will likely apply at least one method, possibly all three, to a given house and then reconcile the respective results to come up with an estimate value. A final step would be to adjust that value up or down, based on market conditions.

Sales comparison is where an appraiser analyzes recent sales of similar homes, generally known as comps. If one of the comps has solar, the appraiser can evaluate the differences between the various homes and their respective sales prices and then identify a contributing value for the PV system. Right now, however, solar is still not widely adopted, so relying on comps isn't always feasible.

The second method is cost. There are a number of ways that an appraiser can use cost to determine value. They generally involve taking the actual cost of the system, identifying when it was installed, and then applying amortization schedules and formulas to come up with a current value figure.

However, there does seem to be a gap with the cost method. It never considers how well the PV system works or whether the system works at all, which may miss solar's true value.

As an analogy, say you struck oil in your backyard. It doesn't make sense that the value of the oil well should

Financial Analysis

be its depreciated cost or the cost of putting another oil well back there. The key figure should be the estimated value of the oil that can be extracted by the well. The same applies with solar. The key figure should be the value of the electricity that the PV system will produce.

This leads to the third appraising method: income. With a PV system, income is the savings that the system gives you. Using this method, you estimate the total savings over the remaining years of the system's useful life and then calculate how much those savings are worth today.

It is the same calculation that is often applied for lottery winners. If you won a gazillion dollars in the lottery, you generally have the option of having your winnings paid to you in increments over thirty years or taking it all at once in a smaller, lump-sum payment. The lump-sum payment is determined by calculating how much money should be invested right now to end up with a gazillion dollars at the end of thirty years.

Solar companies generally use a variation of the income method in their proposals. However, those proposals often provide only one value-add figure, showing a "snapshot in time" of the increased value to your home.

The spreadsheets in the appendices take this further by calculating the estimated value add for each year of the analysis. First, they come up with the Future Savings number by totaling the projected savings that the PV system will generate over its remaining years. Then they use a financial calculation to determine how much those savings are worth today, which they display in the Estimated Value Add column. That is the present value of those future savings for a particular year.

Year	Replace Inverter	Federal Tax Credit	PBIs & Other Incentives	Annual Elec Savings	Net Cash Flow	Running Total	Future Savings (FS)	Estimated Value Add (Present Value of FS)
Year 0					-20,441	-20,441		
Year 1		6,132	194	1,189	7,465	-12,976	42,052	19,312
Year 2				1,224	1,365	-11,611	40,828	19,368
Year 3				1,261				19,388
Year 4				1,298				19,370
Year 5				1,337				19,309
Year 6				1,377				19,202
Year 7				1,418				19,044
Year 8				1,460				18,830
Year 9				1,504				18,555
Year 10				1,549				18,215
Year 11				1,595				17,802
Year 12				1,642				17,311
Year 13				1,691				16,736
Year 14				1,742				16,068
Year 15				1,794				15,300
Year 16				1,847				14,425
Year 17				1,902				13,432
Year 18				1,959	1,845	10,180	15,451	12,314
Year 19				2,018	1,898	12,077	13,434	11,059
Year 20				2,078	1,952	14,029	11,356	9,656
Year 21				2,140	2,007	16,036	9,216	8,095
Year 22				2,204	2,065	18,101	7,013	6,363
Year 23				2,269	2,123	20,224	4,744	4,446
Year 24				2,337	2,184	22,408	2,407	2,330
Year 25				2,407	2,246	24,654	0	0
			1,896	43,241				

The spreadsheet adds projected savings over the remaining years to come up with Future Savings.

It then uses a financial calculation to determine what those Future Savings are worth today...

...that is the estimated value added by the PV system.

Drawing Conclusions

So now we have covered the implications of cash flow, both for purchasing with cash and for financing with a loan. And you can see the potential value that a PV system can add to a home. But how do you use this information to determine whether going solar makes financial sense for you?

Financial Analysis

Cash Option: Drawing Conclusions

When purchasing with cash, you need to analyze two sets of numbers, asking yourself the following questions:

- **Running Total**: At what point will I get back the money that I invested and how well will that investment perform after that?

- **Potential Net Impact**: Does the value added by the PV system exceed the current level of my out-of-pocket expenses? The numbers in the Potential Net Impact column provide the answer to that question as they are the results of adding Running Total (where you're at with your investment) and Estimated Value Add (what the future savings are worth today).

	Annual Elec Savings	Net Cash Flow	Running Total	Future Savings (FS)	Estimated Value Add (Present Value of FS)	Potential Net Impact
Year 0		-20,441	-20,441			
Year 1	1,189	7,465	-12,976	42,052	19,312	6,336
Year 2	1,224	1,365	-11,611	40,828	19,368	7,757
Year 3	1,261	1,398	-10,213	39,567	19,388	9,175
Year 4	1,298	1,432	-8,782	38,269	19,370	10,588
Year 5	1,337	1,466	-7,315	36,932	19,309	11,994
Year 6	1,377	1,502	-5,813	35,555	19,202	13,389

Using those numbers, what conclusions can you draw from the example on the next page? The Running Total will go from negative to positive in year 10, so that is when you will have recouped all of your investment. While you are waiting for that, the positive Potential Net

Making Solar Pay

Impact shows that the estimated value added to your home should outweigh any negative balance in the Running Total.

The numbers to analyze if you use cash to purchase your PV system.

	Annual Elec Savings	Net Cash Flow	Running Total	Future Savings (FS)	Estimated Value Add (Present Value of FS)	Potential Net Impact
Year 0		-19,941	-19,941			
Year 1	1,189	7,315	-12,626	42,052	19,312	6,686
Year 2	1,224	1,365	-11,261	40,828	19,368	8,107
Year 3	1,261	1,398	-9,863			9,525
Year 4	1,298	1,432	-8,432			10,938
Year 5	1,337	1,466	-6,965	Running Total		12,344
Year 6	1,377	1,502	-5,463	goes positive		13,739
Year 7	1,418	1,539	-3,924	in Year 10.		15,120
Year 8	1,460	1,577	-2,346			16,484
Year 9	1,504	1,616	-730			17,826
Year 10	1,549	1,657	927			19,142
Year 11	1,595	1,514	2,441			20,243
Year 12	1,642	1,557	3,998	Potential Net		21,309
Year 13	1,691	1,602	5,600	Impact is		22,335
Year 14	1,742	1,648	7,247	positive		23,315
Year 15	1,794	-2,100	5,147	every year.		20,448
Year 16	1,847	1,744	6,891			21,316
Year 17	1,902	1,794	8,685			22,117
Year 18	1,959	1,845	10,530	15,451	12,314	22,843
Year 19	2,018	1,898	12,427	13,434	11,059	23,486
Year 20	2,078	1,952	14,379	11,356	9,656	24,035
Year 21	2,140	2,007	16,386	9,216	8,095	24,481
Year 22	2,204	2,065	18,451	7,013	6,363	24,814
Year 23	2,269	2,123	20,574	4,744	4,446	25,020
Year 24	2,337	2,184	22,758	2,407	2,330	25,088
Year 25	2,407	2,246	25,004	0	0	25,004
	43,241					

So it comes down to this: If you invest in a PV system, one of two things will happen. One, you stay in your house long term, in which case the system will pay for itself. Or two, you sell the house after installing the PV system, and the value added by the system should enable you to get back more money than you spent.

In this example, using cash to invest in a PV system makes financial sense.

Financial Analysis

Loan Option: Drawing Conclusions

If you are borrowing money to finance a PV system, the numbers and questions that you need to evaluate are as follows:

- **Net Cash Flow**: Am I net positive? Meaning, is the amount that I am saving more than the cost of my finance charges?

- **Potential Net Impact**: Are the accrued savings and added value greater than my loan balance? Again, the spreadsheet answers that question by calculating Potential Net Impact, this time using three inputs:
    - Loan Balance: how much you owe on the loan
    - Running Total: how much you have saved thus far
    - Estimated Value Add: the value of savings yet to come

	Loan Balance	Finance Charges	PBIs & Other Incentives	Annual Elec Savings	Net Cash Flow	Running Total	Future Savings (FS)	Estimated Value Add (Present Value of FS)	Potential Net Impact
Year 0	-13,959				0	0			
Year 1	-13,561	-1,190	194	1,189	143	143	42,052	19,312	5,894
Year 2	-13,142	-1,190	193	1,224	175	318	40,828	19,368	6,543
Year 3	-12,702	-1,190	192	1,261	208	525	39,567	19,388	7,211
Year 4	-12,235	-1,190	191	1,298	242	767	38,269	19,370	7,902
Year 5	-11,739	-1,190	190	1,337	276	1,043	36,932	19,309	8,613

Applying this to the loan example, you can see that you are net positive. You bought the PV system with no money down, the savings produced by the system are

81

covering the cost of the loan, and the loan will be paid off in time. The system saves you money and pays for itself, very slowly but in a methodical, sustainable way.

The numbers to analyze if you finance your PV system with a loan.

	Loan Balance	Finance Charges	Net Cash Flow	Running Total	Future Savings (FS)	Estimated Value Add (Present Value of FS)	Potential Net Impact
Year 0	-13,959		0	0			
Year 1	-13,561	-1,190	143				5,894
Year 2	-13,142	-1,190	175				6,543
Year 3	-12,702	-1,190	208				7,211
Year 4	-12,235	-1,190	242				7,902
Year 5	-11,739	-1,190	276				8,613
Year 6	-11,223	-1,190	312				9,335
Year 7	-10,678	-1,190	349				10,070
Year 8	-10,099	-1,190	387				10,823
Year 9	-9,485	-1,190	426				11,589
Year 10	-8,843	-1,190	467				12,357
Year 11	-8,159	-1,190	324				12,952
Year 12	-7,440	-1,190	367				13,547
Year 13	-6,672	-1,190	412				14,151
Year 14	-5,870	-1,190	458				14,744
Year 15	-5,018	-1,190	-3,290				11,538
Year 16	-4,125	-1,190	554				12,109
Year 17	-3,176	-1,190	604				12,669
Year 18	-2,171	-1,190	655				13,211
Year 19	-1,117	-1,190	708				13,717
Year 20	0	-1,190	762	4,537	11,356	9,656	14,193
Year 21			2,007	6,544	9,216	8,095	14,639
Year 22			2,065	8,609	7,013	6,363	14,972
Year 23			2,123	10,732	4,744	4,446	15,178
Year 24			2,184	12,916	2,407	2,330	15,246
Year 25			2,246	15,162	0	0	15,162

You are net positive in Year 1.

Potential Net Impact is positive every year, so the accrued savings and added value are greater than the loan balance.

Potential Net Impact is positive in year 1 and remains positive throughout the useful life of the PV system. So if you opt to sell your home while you are paying back that loan, the positive impact of the PV system should be higher than your negative loan balance.

In this example, borrowing the money to purchase a PV system makes financial sense.

Seven
Installation

Installation of grid-direct PV system is pretty straight forward. After a contract is signed, your new solar best friend will want to take a closer look at your house to expand on the initial evaluation done for the proposal. You need to be aware of issues that can arise from that closer look.

Common Installation Issues

Before installing a PV system, the solar company will likely conduct an extensive site audit. A key aspect of the audit will be an in-depth analysis of your current electrical design, focusing primarily on how the power from the utility grid is dispersed to the various loads inside your house.

This leads to the most common installation issue: the need to upgrade your electrical service. The electrical service panel may be full (meaning there is no room for additional connections or breakers), or it may not be rated to handle the electric current flowing through the panel once solar is added. Both of those situations may require a larger or a second panel. Even if the service panel is good to go, you might have to reroute existing wires and conduits. And you may have to move the service panel completely if you want to colocate the solar components and there is not sufficient space in the current location.

Electrical upgrades are more complicated if you are installing a battery-backup system. That will require at least one additional subpanel for the essential house loads that the battery will power during a blackout. Another issue is that the battery and its related components generally are not designed to be mounted outside. You will have to carve out some space in your garage, your basement or some other area to place this equipment and the corresponding wires and conduits.

The roof is another common area where work is required. The solar company will take extensive measurements of your roof to ensure the solar design fits within local building codes and adequately factors in obstructions such as chimneys, plumbing vents, AC units, and whatever else. The company may also identify areas that need minor repairs. If your roof is very old, the ideal situation would be to replace it just prior to installing solar since roofs and PV systems have about the same useful life.

Less frequent are structural issues. Mounting a PV array involves a lot of equipment, such as footings, the frame (known as a racking system), electrical conduit, and the modules that make up the array. When it is all said and done, a PV system weighs about three to four pounds per square foot. If you have a 200 square-foot array, you may be adding 800 pounds to your roof. Part of the site audit will be determining that your roof can

Installation

handle the additional stress and recommending steps to shore up support, if necessary.

Another issue involves shade. A large tree can cast shadows that limit the sun's ability to hit the array during certain times of the day and year. If the tree is in your yard, you may be able to temporarily resolve the situation by having the tree trimmed. A permanent solution would be to remove the tree entirely, but there is a bit of irony in that. Perhaps you could justify it by stating you are sacrificing one tree to save many more. You may convince yourself, but I'm not sure your neighbors will buy it.

Fortunately, any spot work to repair a roof or ready the site for solar will likely qualify as an installation expense subject to the 30 percent federal tax credit. However, as mentioned in the chapter on incentives, you should consult with a tax advisor before replacing your roof or taking on a major expense that you plan to claim for the credit.

Other Issues to Consider

Whenever you are doing a home-improvement project, you have to consider the gotchas, the little (and sometimes big) surprises that always seem to pop up. You know, the part of home renovation TV shows when owners and contractors start rolling their eyes and wailing about the added cost. Most gotchas are unforeseen, but there are a couple that you should be aware of so you are not caught off guard.

Probably the most common involves compliance with the local electrical code. Since you are installing an electrical system on your house, the solar company will need to get a permit for the work and that work will be inspected to make sure that it meets local standards. The potential gotcha is you may be flagged for a violation if your house is not currently up to code, and you will have to fix that issue, even if it is not related to the PV system.

For example, say you had an air conditioner replaced a few years ago. If that work does not meet regulations, an electrical inspector may notice that and call out the violation. Certainly, the air conditioner is not related to the PV system, but you would have to address the deficiency before the house would pass the overall electrical inspection.

One way to anticipate a potential problem is to think back on previous work done to your house. The odds of an unpleasant surprise will be higher if you have had any significant additions or renovation projects where permits were not issued. This might be because work was done by you, a friend, or a contractor who decided to cut a few corners. If someone took a shortcut then, it may come back to bite you now.

It's a good idea to discuss past projects with the solar company's electricians and engineers before installation begins. They can help you identify potential issues and possible remedies if you do encounter a problem.

Another potential gotcha is not really an issue with you; it involves the utility's infrastructure. Keep in mind that the electrical service for many neighborhoods, especially older ones, was not designed to have power flow both to and from the houses. The utility installed its equipment with the intent of having power go one way only: from the grid to the individual home. If the utility company has to upgrade its infrastructure, such as replacing a transformer on a pole outside your house, the utility may charge you for all or a portion of that cost, going on the premise that it became an issue only because you wanted to add a net-metering PV system. Thus, you should share in that cost.

While this does not happen often, it is something you should discuss with the solar company. Ask them about what issues could arise from making that final connection to the electrical grid. And ask them how the utility has handled problems related to that in the past.

Eight
Maintenance

This is a very short chapter because the maintenance of a PV system is a very short subject.

A PV system has no moving parts, so the amount of maintenance to keep the system running is limited. The biggest issue that you will likely face is dust or debris collecting on the modules. An excessive amount could affect the array's production. However, given that the modules are generally mounted at angles, they are usually kept clean by rain and snow washing over them.

It is probably a good idea to have a professional service come out every year or two to manually clean the system, remove excessive debris, and tighten any loose connections. The cost for this service should be minimal. If your solar company did a solid installation, then your PV system should remain sound for years to come.

Other than that, you might want to do a visual inspection from time to time, especially after a big storm. This doesn't need to be an involved process. Simply go outside and take a look. You want to see if any modules appear loose or out of place, if debris needs to be cleared away, or if any wires that were hidden are now exposed.

If you do see something, think twice before doing something about it. Just climbing on your roof is a risky proposition. Whatever you do, DO NOT WALK OR CRAWL ACROSS THE ARRAY. Excessive pressure

Making Solar Pay

may weaken or crack a module. Nothing good is going to happen if a tiny, imperceptible crack allows water to seep under the top layer. Compounding that, you may void the warranty covering your system.

In summation, have your PV system cleaned and checked periodically. If you see something that concerns you, call the professionals. They can take care of an issue without causing a bigger problem.

Hey, that's it. We're done. Next chapter.

Nine
Tracking Performance

Your PV system has been installed. The utility company has completed its final commissioning steps. The system is connected to the electrical grid. Now what? How will you know that the PV system is performing as promised?

Electric Bill and Meter

You started this journey by estimating your electric usage and determining your solar goals. Let's use your new electric bill as a scorecard on whether you achieved those goals. An example of a net-metered electric bill is shown on the next page, and you gotta love those zeros!

So what can we tell from this bill? The PV system sent out 1,288 kWh over the course of the month, but the net draw from the utility was zero. It also shows a total energy use for the month as 208 kWh.

The most likely scenario is that 1,288 kWh was sent out to the grid as a net-metered credit during the days, when production was high and usage was low. In the evenings, when there was no production, the first 1,288 kWh pulled from the grid brought the homeowner's net energy usage back up to zero, as prescribed by the typical net-metering agreement. They then used an additional 208 kWh, but the net energy delivered by the utility was still zero. Since the utility provide no billable kilowatt-hours, the remaining 208 kWh were likely pulled from a built-up credit from months when energy sent to the grid exceeded the amount of energy used.

Net usage

ELECTRICITY SERVICE DETAILS

Meter Reading Information

Read Dates: 12/30/15 - 01/28/16 (29 days)

Description	Current Reading	Previous Reading	Usage
Total Energy	98771 Actual	98563 Actual	208 kWh
Net Delivered by Xcel	0 Actual	Actual	0 kWh
Net Generated by Customer	1288 Actual	Actual	1288 kWh

ELECTRICITY CHARGES   **RATE: R Residential General**

Description	Usage Units	Rate	Charge
Service & Facility			$6.75
Non-Summer	0 kWh	$0.046040	$0.00
Trans Cost Adj	0 kWh	$0.000630	$0.00
Elec Commodity Adj	0 kWh	$0.027810	$0.00
Demand Side Mgmt Cost	0 kWh	$0.001220	$0.00
Purch Cap Cost Adj	0 kWh	$0.006500	$0.00
CACJA	0 kWh	$0.003920	$0.00
Renew Energy Std Adj			$0.14
GRSA			$0.63
Subtotal			**$7.52**
Franchise Fee		3.00%	$0.23
Sales Tax			$0.28
Total			**$8.03**

It is important to note that the bill only gives a view of consumption. It says nothing about your PV system's overall production. It does not account for the electricity that you used prior to sending out excess energy to the grid. One thing is for certain, though: the system is doing exactly what it supposed to do. The total charge for the month was $8.03!

Tracking Performance

Of course, you have to wait until the electric bill arrives at the end of the month before you can analyze it. What if you want to know right now whether your PV system is working? For that, you can use the new electric meter.

When commissioning your PV system, the utility company will install a meter designed to track energy going to and from your house. This new "net" meter can tell you whether you are drawing energy from the grid or sending it out. In the example, the kWh bars increase incrementally when you are pulling from the grid, going from one bar to two bars to three bars, moving from left to right.

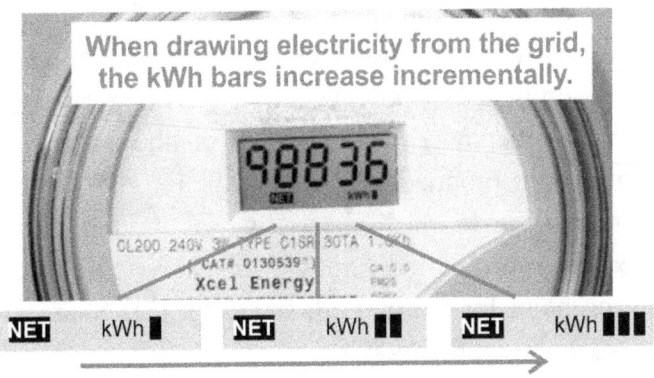

When you are sending energy out, the bars move in the opposite direction (going backward, if you will), decreasing from three bars to two to one. The obvious time to check for this would be during the middle of a bright, sunny day. It is during this peak sun hour period that the PV system should be generating more energy than the house is using.

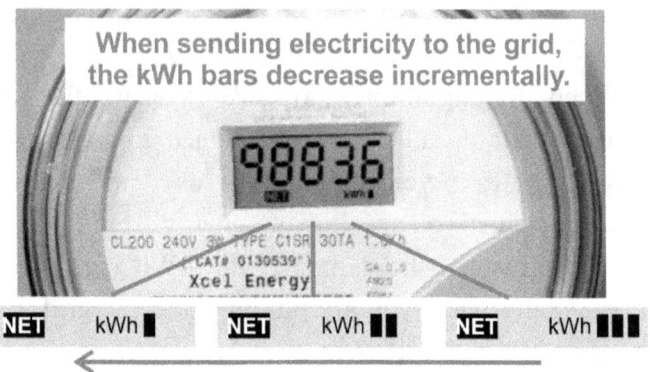

Some utility companies require an additional electric meter, one dedicated to tracking solar production. If a production meter is showing activity, then the PV system is generating power. The numbers on the meter would show overall production since installation, so you could not identify how much electricity was produced over a specified time period unless you knew the meter's starting point for that period, such as a particular day or the beginning of the month.

Monitoring Production

But what if you do want details about production numbers? How do you get those? Many inverters have a built-in display that shows production data and other

Tracking Performance

statistics. However, this data can be very technical and difficult to read.

For more easily digestible information, most inverters have monitoring capabilities accessible from a computer, tablet, or smartphone. These can convey production levels for several time frames, shown in a variety of views. One common one is a snapshot overview.

Overview			
Current Power	Energy today	Energy this month	Lifetime energy
4.62 kW	**26.29 kWh**	**788.7 kWh**	**17,351 kWh**

Another common view is the bar chart. Most monitoring features provide several types. The chart below shows production for one day on an hourly basis. The one on the next page tracks production over the course of a month.

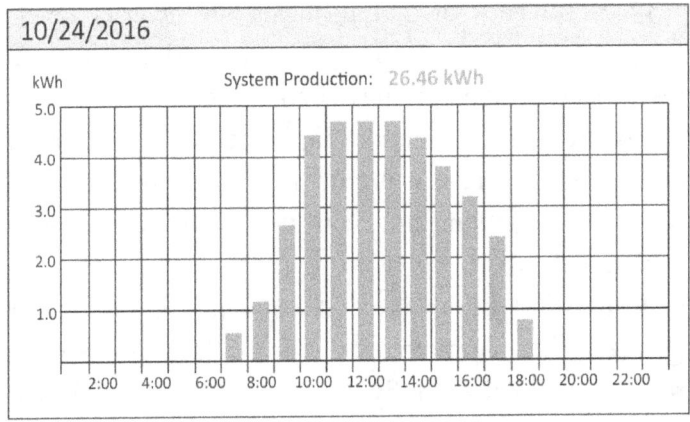

Making Solar Pay

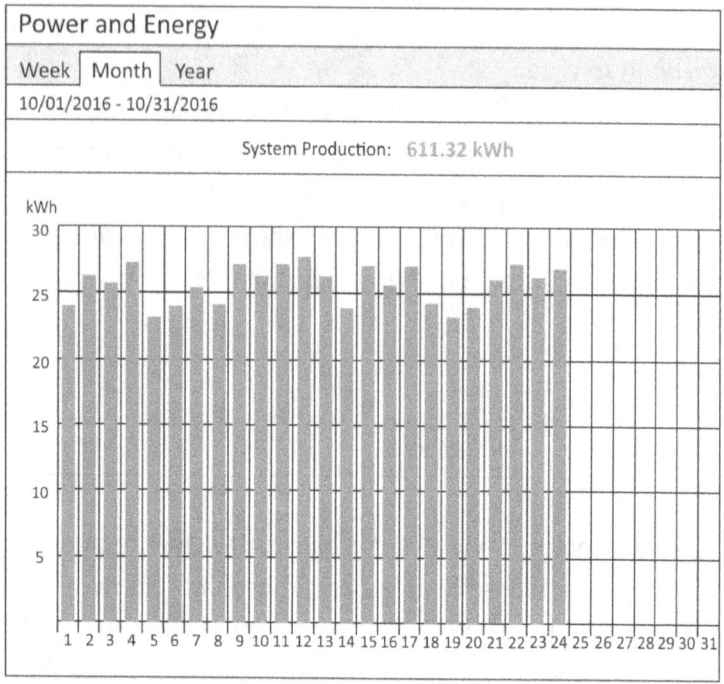

One drawback of monitoring at the inverter level is that you can only view the system as a whole. Since the power from all the modules is combined before going to the inverter, the production data that you see is for the entire unit. If there is a problem, you will be aware of a drop in production, but you won't know where the problem is.

That won't be an issue if you have a system with optimizers or microinverters. Those allow you to track production down to individual modules. The following images show the layout of an array and the kilowatt-hour production for each module.

Tracking Performance

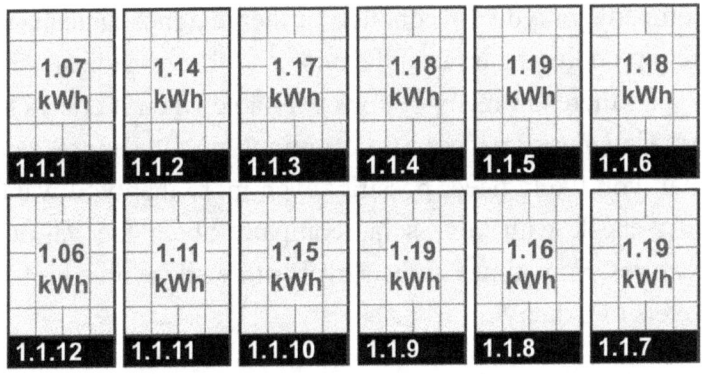

As you can see, you will be able to pinpoint a troublesome module if that module's production drops off or shuts down completely.

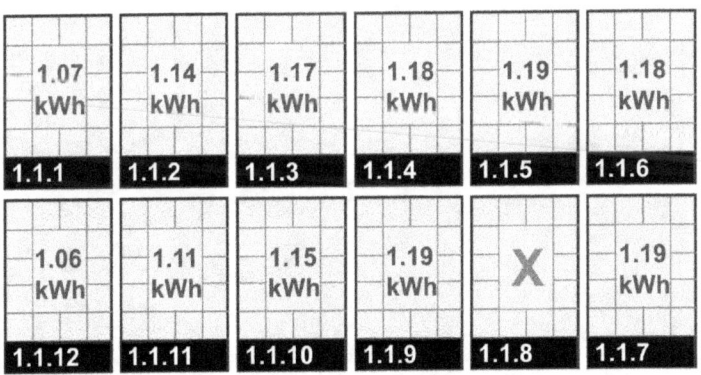

Web-based applications often allow you to set alerts to notify you with an e-mail or text message if production drops below a certain level. That way you do not have to constantly monitor your PV system. Your system will monitor itself and let you know if there's a problem. Often the monitoring feature then enables your solar company or you to conduct further diagnostic tests to help pinpoint the issue and devise possible remedies.

So there are many options available when it comes to tracking performance. Some options provide more data than you likely need, possibly even more than you want. Just check with your solar company about the various features. Sometimes higher levels of detail involve higher levels of expense.

Ten
Case Study: How One Homeowner Used This Book to Determine Whether Solar Makes Financial Sense

What follows is how one homeowner applied the theory and techniques presented in this book to analyze a potential solar energy system. Each section includes a reference to the chapter or part of the book that covers that section's topic. The case itself is taken from actual events, although the homeowner's real name is not David.

SPOILER ALERT: David did opt to install a PV system, which probably won't come as a surprise. But the conclusion that he ultimately reached does provide an interesting perspective on his solar experience.

Gathering Data (Chapter 1)

David knew that I was a solar guy and said he wanted to explore adding a PV system to his home. I told him that I was writing a book on that, and he agreed to test the process that I was developing.

The first step would be to gather data on his electrical usage. David had lived in his home for several years, so

Making Solar Pay

he had plenty of data available. Since he is the type who saves all his statements and receipts, he easily gathered his utility bills for the past twelve months. He then entered the data from those bills into the electrical-usage spreadsheet.

Electrical Usage and Cost

Past 12 Months--kWh Usage and Cost

	Month	kWh used	Cost
1	April 2016	688	78.64
2	March 2016	821	94.58
3	February 2016	699	81.78
4	January 2016	801	96.44
5	December 2015	604	80.82
6	November 2015	510	69.51
7	October 2015	610	84.18
8	September 2015	954	118.87
9	August 2015	1,102	128.93
10	July 2015	931	108.93
11	June 2015	805	93.14
12	May 2015	676	77.94
	Totals	9,201	1,113.76
	Cost per kWh		0.1210

From this, David determined two key bits of information:

- Total usage: 9,201 kWh
- Cost per kWh: 0.1210

He contacted three solar companies and gave them the usage number. He also told them his goal was to

Case Study

completely offset that usage and the annual electrical expense.

Comparing Proposals (Chapter 5)

The three solar companies provided David with custom proposals. David pulled the following information from each:

Company 1
- Proposed system size: 6.2 kW
- Estimated production: 9,275 kWh
- Cost: $20,150

Company 2
- Proposed system size: 6.3 kW
- Estimated production: 9,156 kWh
- Cost: $19,975

Company 3
- Proposed system size: 6.25 kW
- Estimated production: 9,005 kWh
- Cost: $21,810

David used that data to complete the inputs on the Proposal Comparison spreadsheet.

Making Solar Pay

David entered size, production, and cost inputs for the three companies that he was analyzing.

Proposal Comparison

Goal	9,201		Number of companies bidding	3	

First Cut: Analyze Cost

	Company 1	Company 2	Company 3	Company 4	Company 5
System					
→ Proposed size (kW)	6.20	6.30	6.25	1.00	1.00
Average peak sun hours	4.10	3.98	3.95	0.00	0.00
→ Estimated production	9,275	9,156	9,005	1	1
% offset	100.8%	99.5%	97.9%	0.0%	0.0%
Cost					
→ Cost	20,150	19,975	21,810	1	1
Cost per watt	3.25	3.17	3.49	0.00	0.00

He glanced at the figures for average peak sun hours. Seeing that those figures were about the same for all three companies, he knew that he could make a fair comparison of the proposals.

Proposal Comparison

Goal	9,201		Number of companies bidding	3	

First Cut: Analyze Cost

	Company 1	Company 2	Company 3	Company 4	Company 5
System					
Proposed size (kW)	6.20	6.30	6.25	1.00	1.00
Average peak sun hours	4.10	3.98	3.95	0.00	0.00
Estimated production	9,275	9,156	9,005	1	1
% offset	100.8%	99.5%	97.9%	0.0%	0.0%
Cost					
Cost	20,150	19,975	21,810	1	1
Cost per watt	3.25	3.17	3.49	0.00	0.00

For the first cut, he compared each company's cost per watt and eliminated the highest one.

Case Study

Proposal Comparison

	Company 1	Company 2	Company 3	Company 4	Company 5
Goal	9,201		Number of companies bidding	3	

First Cut: Analyze Cost

	Company 1	Company 2	Company 3	Company 4	Company 5
System					
Proposed size (kW)	6.20	6.30	6.25	1.00	1.00
Average peak sun hours	4.10	3.98	3.95	0.00	0.00
Estimated production	9,275	9,156	9,005	1	1
% offset	100.8%	99.5%	97.9%	0.0%	0.0%
Cost					
Cost	20,150	19,975	21,810	1	1
Cost per watt	3.25	3.17	3.49	0.00	0.00

For a second cut, David used the lower portion of the spreadsheet, which levels the production estimates by applying the same number for "average peak sun hours." This affects a company's estimated production and % offset. He wanted to see if a modified estimate and offset would make a company's cost per watt seem more or less attractive.

Second Cut: Level Production Estimates

	Company 1	Company 2	~~Company 3~~	Company 4	Company 5
System					
Proposed size (kW)	6.20	6.30	~~6.25~~	1.00	1.00
Average peak sun hours	4.01	4.01	~~4.01~~	4.01	4.01
Estimated production	9,077	9,223	~~9,150~~	0	0
% offset	98.7%	100.2%	~~99.4%~~	1.0%	1.0%
Cost					
Cost	20,150	19,975	~~21,810~~	1	1
Cost per watt	3.25	3.17	~~3.49~~	0.00	0.00

He found that Company 2 came closer to meeting his goal and at a lower cost. He chose Company 2.

Financial Analysis (Chapter 6 and Various Appendices)

Now David needed to verify that Company 2's proposal made financial sense. Interest rates had recently dropped, which led him to think about refinancing his home. That gave him the option of adding the cost of solar to a new mortgage.

- Original thirty-year mortgage
  - Balance of current loan: $324,000
  - Original interest rate: 4.75%
  - Original monthly payment: $1,690

- Refinanced thirty-year mortgage
  - Size of the new loan (adding $20,000 for solar): $344,000
  - New interest rate: 4.25%
  - New monthly payment: $1,692
  - New monthly payment if solar was not included in the loan: $1,594
  - Solar's impact on the new payment: $98 ($1,692 - $1,594)

He used the $98 solar impact as an input for the financial-analysis spreadsheet discussed in appendix D, the one that incorporates solar into a mortgage. After entering all the inputs, David focused on the two sets of numbers that drive a mortgage analysis:

Case Study

- **Net Cash Flow**: If you are net positive, the amount that you save each year is greater than the amount of solar's added cost to the loan.

- **Potential Net Impact**: A positive number means that the amount you have saved thus far and potential value added by the PV system is greater than the amount that you owe on the loan.

Looking at the spreadsheet shown on the next page, David saw that he would be net positive in year 1. He could buy the PV system with no money down and the savings produced by the system would cover solar's portion of the monthly payment.

He also saw that the Potential Net Impact would be positive in year 1 and would remain positive throughout the life of the PV system. So if he ever sold his home, the positive impact of the PV system should be higher than his negative loan balance.

David concluded that adding the cost of solar to his new mortgage was a good investment. He completed the refinance, and he bought the PV system.

David then applied his own twist to the analysis. This book advocates comparing two new loan payments: one that includes solar and one that does not. The difference is solar's monthly impact on the mortgage.

David took it a different direction. He compared his new solar-inclusive payment with what his payment had

Making Solar Pay

been with the original mortgage. The new payment was $1,692; the old payment was $1,690, a difference of $2.

"My mortgage payment really didn't change," David said. "I'm essentially paying the same amount that I was before, but now I have a PV system. So as far as I'm concerned, I got solar for free."

	Loan Balance	Annual Elec Savings	Net Cash Flow	Running Total	Future Savings (FS)	Estimated Value Add (Present Value of FS)	Potential Net Impact
Year 0	-19,975		0	0			
Year 1	-19,675	1,108	6,058	6,058	39,192	17,164	3,547
Year 2	-19,366	1,141	95				4,034
Year 3	-19,036	1,175	125	David is net positive, so his savings are greater than the cost of servicing the loan.			4,542
Year 4	-18,697	1,210	157				5,056
Year 5	-18,337	1,246	189				5,584
Year 6	-17,958	1,283	222				6,124
Year 7	-17,558	1,322	256				6,671
Year 8	-17,139	1,361	291				7,223
Year 9	-16,699	1,402	328				7,776
Year 10	-16,240	1,443	365				8,325
Year 11	-15,750	1,486	229	A positive Potential Net Impact means the savings and added value are greater than the loan balance.			8,702
Year 12	-15,241	1,531	269				9,067
Year 13	-14,702	1,576	311				9,424
Year 14	-14,142	1,623	353				9,758
Year 15	-13,543	1,672	-3,068				6,617
Year 16	-12,924	1,722	442				6,904
Year 17	-12,265	1,773	488				7,167
Year 18	-11,576	1,826	535				7,388
Year 19	-10,856	1,880	584				7,557
Year 20	-10,097	1,936	634				7,677
Year 21	-9,298	1,994	686	9,549	8,589	7,485	7,735
Year 22	-8,459	2,054	739	10,287	6,536	5,895	7,723
Year 23	-7,581	2,115	793	11,080	4,421	4,127	7,626
Year 24	-6,652	2,178	849	11,928	2,243	2,167	7,444
Year 25	-5,673	2,243	906	12,834	0	0	7,161

Appendices

Appendix A
Other Types of PV Systems

Chapter 2 talked about two PV systems: grid direct and grid direct with battery backup. Besides those, there are other types of PV systems that can be used for residential sites. While these systems are not available or feasible for all homeowners, it is possible that one of them may be a good option for you.

Stand-Alone System (Off Grid)

As the name implies, a stand-alone system is a PV system that is not connected to the utility grid, frequently referred to as being off grid.

Off-grid systems make the most sense for homes located far from the utility grid, where running and maintaining power lines can be cost prohibitive. Their design is similar to a grid-direct system with battery backup, with the obvious exceptions of not needing the components required for a grid connection. During the day, the array charges the battery and powers the house loads. At night, power from the battery flows to the inverter and then into the house. There is no need for a meter as there is no connection to the grid.

Properly sizing the battery is critical as it has to service all the electrical loads in the house when the sun is not shining. As with any battery system, you are adding additional components and complexity such as devices to regulate charging and discharging. Many off-grid systems also have some kind of backup generator, often one that runs on fossil fuel. That is not the best option if you want to be completely green, but sometimes you don't have a choice. You have a stand-alone system. You're off grid. You are on your own.

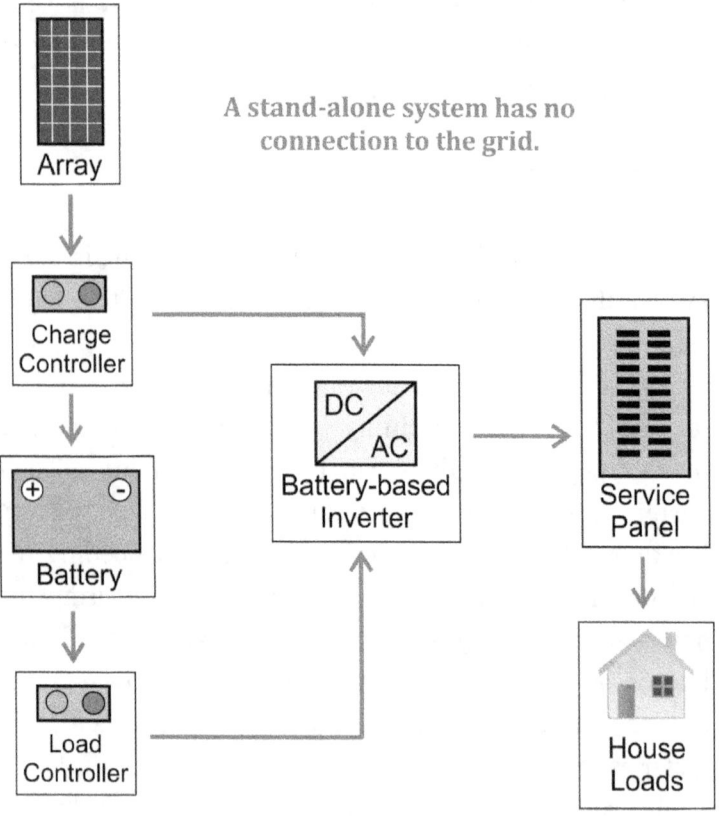

A stand-alone system has no connection to the grid.

Community Solar Gardens

A community solar garden is a very large PV system where all the electricity produced by the system flows onto the utility's grid. While the solar garden is not physically connected to any homes, people can purchase the right to receive some of the credits derived from the energy sold to the utility company. For example, if the garden consists of a 50 kW system, you might have ten different households purchasing 5 kW each. Those households would be able to tie their home accounts with their solar-garden accounts. Then the cost of the electricity they purchase from the utility is offset by their portions of the credits generated by the solar garden.

There are several attractive features with this setup:

- No solar equipment needs to be installed at your home.

- Solar becomes viable to people who would not qualify otherwise, such as those who don't own their homes or those whose homes are not good solar candidates.

- Solar credits may be transferrable, so you should be able to sell your stake in the garden, give it to someone else or a charity, or possibly even take the service to a new address if you move.

Making Solar Pay

The primary downside is that a community solar garden will not impact your home's value. Since no equipment is located at your home, the solar electric system cannot be included in a real estate appraisal. As we saw in the chapter on financial analysis, that can have a huge impact on the financial viability of a PV system.

Zero-Sell System

A zero-sell system is similar to a grid-direct system with battery backup. Just like a battery-backup system, the array for a zero-sell system generates electricity for the house and charges the battery during the day. Unlike a battery-backup system, however, the battery is the primary source of power at night. No electricity is ever sent (or sold) to the utility grid. As such, the grid acts much like a backup source, providing electricity for the house when it needs more power than can be supplied by the array and battery.

A zero-sell system may be a good option if there are issues with connecting a PV system to the grid for net metering. Perhaps infrastructure obstacles in your area prevent a standard grid-direct connection. Maybe the utility company has reached capacity for handling PV-generated electricity. Or it might simply be a matter that the utility does not offer net metering or its net-metering program does not make sense for you financially.

Other Types of PV Systems

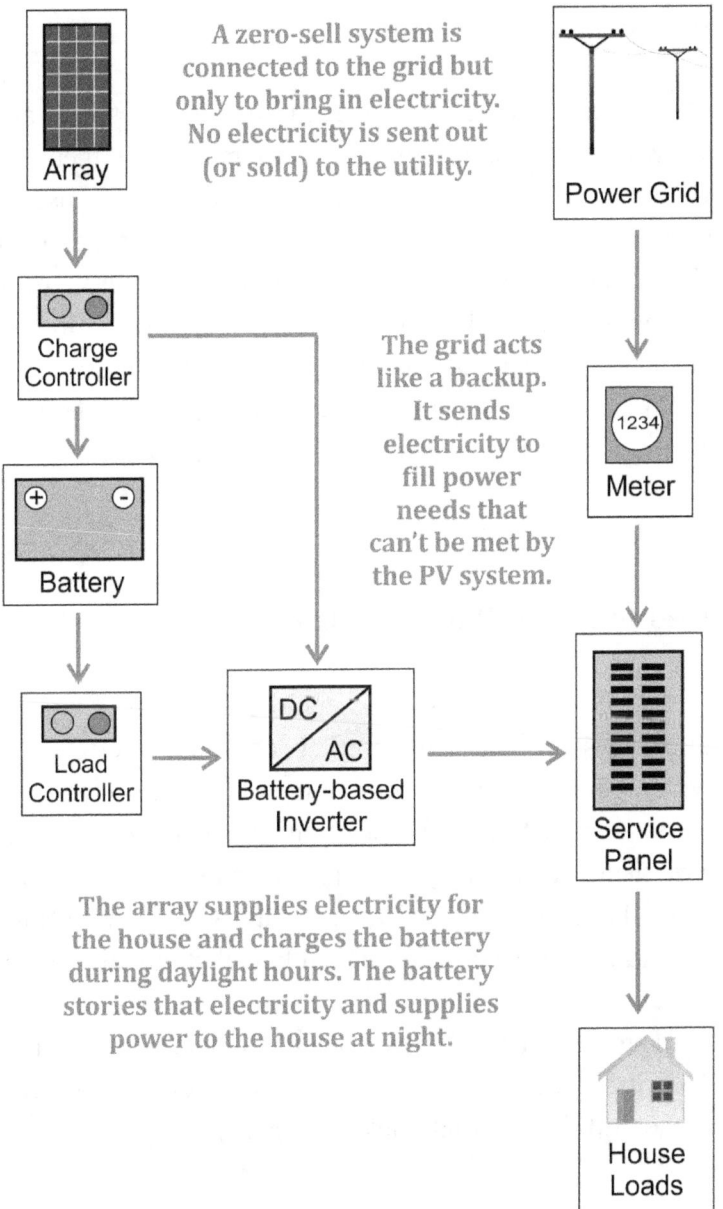

DC Direct

Self-contained devices that run on DC electricity are known as DC-direct systems. You might see them as water pumps, aerating fountains, or vent fans, just to name a few devices. These are the simplest applications of solar energy. Because they run on DC electricity, there are no inverters. The module is wired directly to the device it is powering, often times mounted right on top of it. So sunlight hits the module; the module provides DC electricity to the device; the device runs and runs and runs.

Feed-In Tariffs (Buy All / Sell All)

A feed-in tariff system is another variation of a grid-direct system. Instead of first consuming electricity within your house and then sending the excess to the grid, the system pushes all the electricity generated to the grid. You then draw all the power that you need for your home from the grid, just as you do today. Given this setup, feed-in tariff systems are frequently called buy-all/sell-all systems.

The system needs two meters to function properly. The first tracks all the energy that you produce and send to the grid. The second tracks the energy that you use.

Other Types of PV Systems

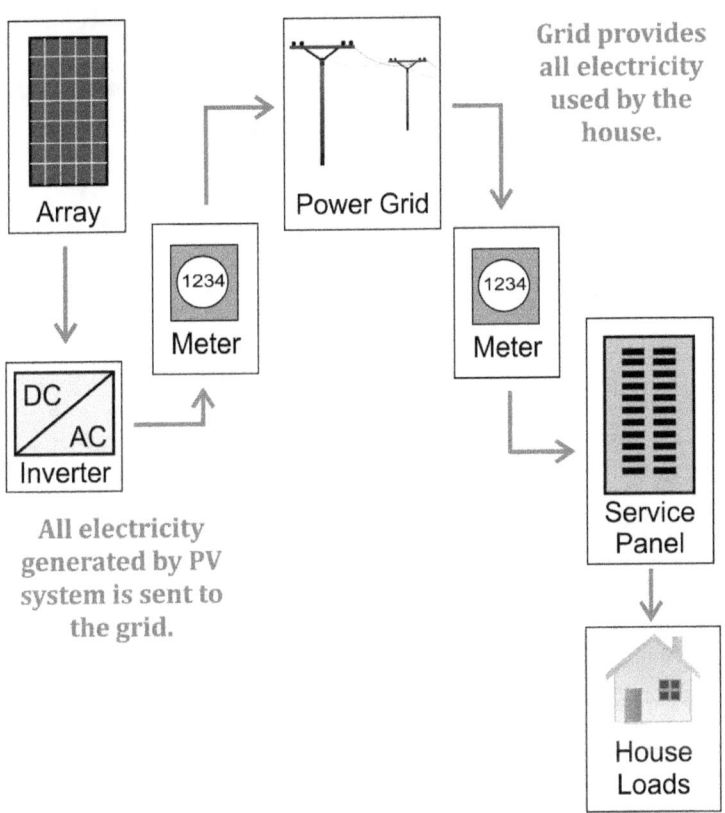

Appendix B
Spreadsheet: Cash Option

When to Use This Spreadsheet

This spreadsheet, available at the Mountain Edge Publishing website (www.mepub.com), covers the cash purchase of a PV system.

Spreadsheet Inputs

There are only six spreadsheet inputs, shown as fields with highlighted backgrounds. Four of these inputs are numbers that you have already worked with in previous chapters.

Financial Analysis--Cash Option

System size (in kW)	6.90		Estimated production	9,696
Cost per watt	2.89		System cost	19,941
System degradation rate	0.5%	% loss of production per year	Site prep expenses	0
Base utility elec rate	0.1226	per kWh		
Utility inflation rate	3.50%	% increase in utility elec rate per year		
Maintenance base cost	0.25%	% of system cost to set Year 1 maint		
Maint inflation rate	5.00%	% increase in maint cost per year		
Inverter replace cost	0.55	per watt, incurred in Year 15		
Federal tax credit	30%		FNMA 30-year fixed rate (90-day)	3.30%

- Inputs from the Proposal Comparison spreadsheet (used in chapter 5):
    - **System size (in kW)**
    - **System cost**
    - **Estimated production**

117

Making Solar Pay

Proposal Comparison

From the Proposal Comparison spreadsheet, use the proposed size, estimated production, and cost numbers for the company that you are analyzing.

Goal	9,674					
First Cut: Analyze Cost						
		Company 1	Company 2	Company 3	Company 4	Company 5
System						
Proposed size (kW)		6.75	6.90	6.95	6.80	6.85
Average peak sun hours		3.95	3.83	3.82	3.85	3.82
Estimated production		9,725	9,638	9,699	9,545	9,555
% offset		100.5%	99.6%	100.3%	98.7%	98.8%
Cost						
Cost		20,567	19,941	23,226	23,868	20,148
Cost per watt		3.05	2.89	3.34	3.51	2.94

- Input from the utility's website or from the Electrical Usage and Rate spreadsheet (used in chapter 1):
    - **Base utility elec rate**

	Electrical Usage and Cost		
	Past 12 Months--kWh Usage and Cost		
	Month	kWh used	Cost
1	March 2016	688	74.53
2	February 2016	636	69.47
3	January 2016	763	84.79
4	December 2015	677	78.01
5	November 2015	588	68.82
6	October 2015	688	82.84
7	September 2015	1,042	139.38
8	August 2015	1,180	160.84
9	July 2015	1,009	139.22
10	June 2015	883	109.99
11	May 2015	754	88.21
12	April 2015	766	89.66
	Totals	9,674	1,185.76
		Cost per kWh	0.1226

Enter your cost per kWh on the financial spreadsheet as the base utility elec rate.

Cash Option

- Other inputs
    - **Site prep expenses**: Enter the estimated cost for any work that needs to be done to prepare your house for the PV installation. Hopefully, that cost will be zero.

    - **FNMA 30-year fixed rate (90-day)**: This interest rate is often used in calculating the value add of a solar PV system. The current rate is posted on the Fannie Mae website. The easiest way to get there is to do an online search for "FNMA 30 year fixed rate." You will likely get a couple FNMA links in your search results. Either one will work. Just make sure that website you are going to is fanniemae.com.

```
April 2016 30 Year Fixed - Fannie Mae
https://www.fanniemae.com/content/datagrid/hist_net_yields/cur30.html ▼ Fannie Mae
30-YEAR FIXED RATE AT A. DATE, TIME, 10-DAY, 30-DAY, 60-DAY, 90-DAY. 08/01/2016, 08 15
02.78815, 02.80903, 02.84071, 02.87483. 08/02/2016, 08.15

Historical Daily Required Net Yields - Fannie Mae
https://www.fanniemae.com/ /historical-daily-required-net-yields ▼ Fannie Mae
Historical daily required net yields for 10-, 30-, 60-, and 90-day mandatory delivery whole loan
commitments for 30- and 15-year fixed-rate mortgages (FRMs) with Actual/Actual (A/A) remittance
are available by month for the ... Mortgage Type
```

The link with the month in the title will take you directly to the page with the current rates. If you go the historical data route to the Fannie Mae site, click on "30 Year Fixed" for the current month.

Making Solar Pay

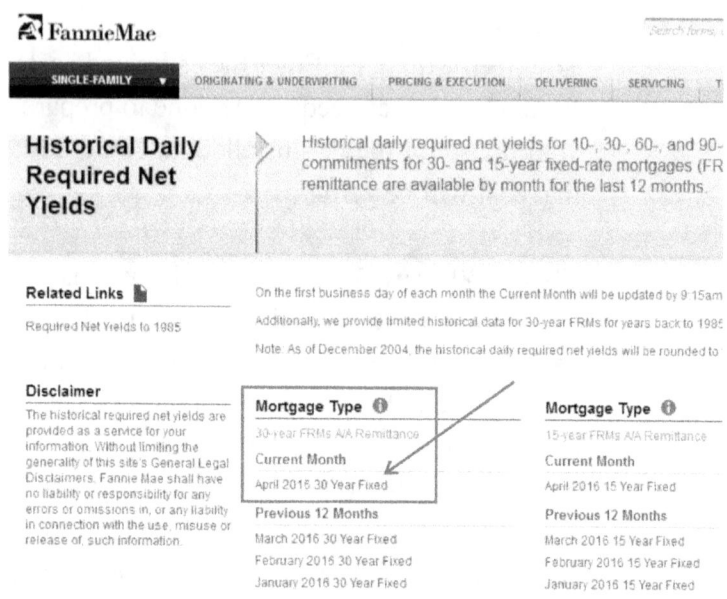

Once you are on the page with the current rates, find the most recent 90-day rate and enter that number in the variable field on the spreadsheet. (The number should display as 3.29%. Put the decimal point in front of the entire number if it reads 329.55% in your spreadsheet.)

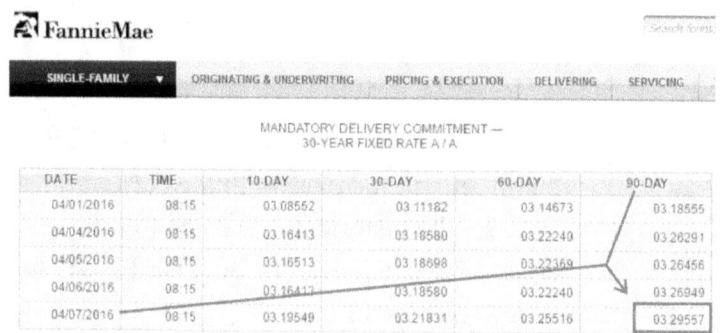

Cash Option

There may be a seventh input, if you are fortunate enough to have it. This is a performance-based incentive (PBI). The company's proposal will likely identify what programs are available in your area and their impact. If one or more apply, enter them in the PBIs & Other Incentives column. For example, a local utility currently has a 0.02-per-kWh incentive for the first ten years of production. So the numbers shown on the spreadsheet are the results of multiplying the production estimate for a given year by 0.02.

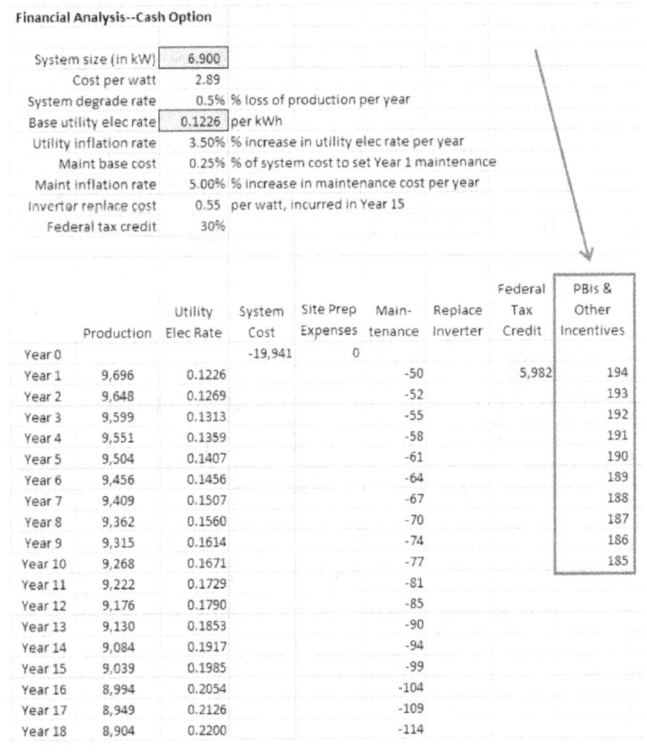

							Federal	PBIs &
		Utility	System	Site Prep	Main-	Replace	Tax	Other
	Production	Elec Rate	Cost	Expenses	tenance	Inverter	Credit	Incentives
Year 0			-19,941	0				
Year 1	9,696	0.1226			-50		5,982	194
Year 2	9,648	0.1269			-52			193
Year 3	9,599	0.1313			-55			192
Year 4	9,551	0.1359			-58			191
Year 5	9,504	0.1407			-61			190
Year 6	9,456	0.1456			-64			189
Year 7	9,409	0.1507			-67			188
Year 8	9,362	0.1560			-70			187
Year 9	9,315	0.1614			-74			186
Year 10	9,268	0.1671			-77			185
Year 11	9,222	0.1729			-81			
Year 12	9,176	0.1790			-85			
Year 13	9,130	0.1853			-90			
Year 14	9,084	0.1917			-94			
Year 15	9,039	0.1985			-99			
Year 16	8,994	0.2054			-104			
Year 17	8,949	0.2126			-109			
Year 18	8,904	0.2200			-114			

(Financial Analysis--Cash Option inputs: System size (in kW) 6.900; Cost per watt 2.89; System degrade rate 0.5% % loss of production per year; Base utility elec rate 0.1226 per kWh; Utility inflation rate 3.50% % increase in utility elec rate per year; Maint base cost 0.25% % of system cost to set Year 1 maintenance; Maint inflation rate 5.00% % increase in maintenance cost per year; Inverter replace cost 0.55 per watt, incurred in Year 15; Federal tax credit 30%)

Spreadsheet Columns

The spreadsheet uses those inputs and various calculations to populate data fields for the following columns:

- **Year**: goes from year 0, which is the year that the PV system is installed, to year 25 so that the spreadsheet can give you data for the normal lifespan of a PV system.

- **Production**: shows the annual production of kilowatt-hours. Production for year 1 is the number that you entered as the estimated production input. Production for year 2 and beyond assumes an annual degradation rate of one half of 1 percent, reached by multiplying the previous year's production number by 0.995.

- **Utility Elec Rate**: shows the rate charged by your utility company for 1 kilowatt-hour of electricity. The year 1 rate comes from what you are currently paying for electricity, as determined in chapter 1 and entered as an input for this spreadsheet. The rates for year 2 and beyond assume an increase each year of 3.5 percent, shown as the default utility-inflation-rate input. In some areas, this escalation will be higher, often

Cash Option

going into a range of 5.0 to 6.0 percent per year. You can update that input as needed.

- **System Cost**: the amount that you are paying the solar company to install and commission your PV system.

- **Site Prep Expenses**: any out-of-pocket expenses that you did not pay to the solar company.

- **Maintenance**: There is very little maintenance with a PV system. The year 1 rate is set as one quarter of one percent, multiplying the system cost by 0.0025. Year 2 and beyond assumes a 5 percent increase over the previous year's maintenance cost.

- **Replace Inverter**: assumes that you will have to replace the inverter around year 15, with the cost being $0.55 per watt. That number is one of the default inputs at the top of the spreadsheet. You might want to check with your solar company for a more accurate number, which you can use to update the inverter-replacement-cost input. Also, if you are using microinverters, ask the solar company about the best way to account for those replacement costs.

- **Federal Tax Credit**: This is the 30 percent tax credit, so the spreadsheet adds System Cost and Installation Expenses and then multiplies that number by 0.30.

- **PBIs & Other Incentives**: the performance-based incentives. Your solar company can identify these for you, if they are available.

- **Annual Elec Savings**: a simple calculation of multiplying production by the utility electrical rate. These savings should rise each year because the rate charged by the utility will likely rise faster than the production level will fall.

- **Net Cash Flow**: gives the financial result for a given year by adding all the revenue and expenses for that time period.

- **Running Total**: shows where you are with your investment by adding the current net cash flow number to the previous year's running total.

- **Future Savings (FS)**: totals the Annual Electric Savings for the PV system's remaining productive years. For example, the Future Saving for year 1 is the total of all the annual savings from year 2 through year 25. For year 2, the number is the annual savings from year 3 to year 25, and so on.

Cash Option

- **Estimated Value Add (Present Value of FS)**: uses the current FNMA 30-year fixed interest rate to calculate how much Future Savings (FS) are worth today.

- **Potential Net Impact**: shows the overall financial picture by adding Running Total to Estimated Value Add.

What to Look For

As stated in the financial analysis chapter, the two numbers to watch are:

- **Running Total**: shows where you are with solar investment. When the number goes positive, you have recovered all of your out-of-pocket costs.

- **Potential Net Impact**: compares the current level of your investment with the estimated value added to your home. A positive number means the overall financial picture is favorable.

In the example on the next page, Running Total goes from negative to positive in year 10, showing when you will have recouped all the money that you invested. While you are waiting for that, the positive Potential Net Impact means that the estimated value added to your home could very well outweigh any negative balance in the Running Total.

Making Solar Pay

The numbers to analyze if you use cash to purchase your PV system.

	Annual Elec Savings	Net Cash Flow	Running Total	Future Savings (FS)	Estimated Value Add (Present Value of FS)	Potential Net Impact
Year 0		-19,941	-19,941			
Year 1	1,189	7,315	-12,626	42,052	19,312	6,686
Year 2	1,224	1,365	-11,261	40,828	19,368	8,107
Year 3	1,261	1,398	-9,863			9,525
Year 4	1,298	1,432	-8,432			10,938
Year 5	1,337	1,466	-6,965	Running Total goes positive in Year 10.		12,344
Year 6	1,377	1,502	-5,463			13,739
Year 7	1,418	1,539	-3,924			15,120
Year 8	1,460	1,577	-2,346			16,484
Year 9	1,504	1,616	-730			17,826
Year 10	1,549	1,657	927			19,142
Year 11	1,595	1,514	2,441			20,243
Year 12	1,642	1,557	3,998	Potential Net Impact is positive every year.		21,309
Year 13	1,691	1,602	5,600			22,335
Year 14	1,742	1,648	7,247			23,315
Year 15	1,794	-2,100	5,147			20,448
Year 16	1,847	1,744	6,891			21,316
Year 17	1,902	1,794	8,685			22,117
Year 18	1,959	1,845	10,530	15,451	12,314	22,843
Year 19	2,018	1,898	12,427	13,434	11,059	23,486
Year 20	2,078	1,952	14,379	11,356	9,656	24,035
Year 21	2,140	2,007	16,386	9,216	8,095	24,481
Year 22	2,204	2,065	18,451	7,013	6,363	24,814
Year 23	2,269	2,123	20,574	4,744	4,446	25,020
Year 24	2,337	2,184	22,758	2,407	2,330	25,088
Year 25	2,407	2,246	25,004	0	0	25,004
	43,241					

In the example above, Running Total goes from negative to positive in year 10, showing when you will have recouped all the money that you invested. While you are waiting for that, the positive Potential Net Impact means that the estimated value added to your home could very well outweigh any negative balance in the Running Total.

So it comes down to this: If you invest in a PV system, one of two things will happen. One, you stay in

Cash Option

your house long term, in which case the system will pay for itself. Or two, you sell the house after installing the PV system, and the value added by the system may enable you to get back more money than you spent.

In this case, using cash to invest in a PV system makes financial sense.

Appendix C
Spreadsheet: Solar Loan Option

When to Use This Spreadsheet

The Solar Loan Option is for when you take out a loan specifically to pay for the PV system. Generally these loans have higher interest rates and shorter terms than mortgages. The example spreadsheet in this appendix is set for a twenty-year term, with several calculations being based on that time frame. The Mountain Edge Publishing website (www.mepub.com) also has spreadsheets for ten, fifteen, and twenty-five year scenarios.

Spreadsheet Inputs

The Solar Loan spreadsheet uses the same inputs as the Cash Option spreadsheet (see appendix B) plus four additional ones:

Financial Analysis--Solar Loan (20 years)

System size (in kW)	6.90		Estimated production	9,696
Cost per watt	2.89		System cost	19,941
System degrade rate	0.5%	% loss of production per year	Site prep expenses	0
Base utility elec rate	0.1226	per kWh		
Utility inflation rate	3.50%	% increase in elec rate per year	Cash down	0
Maintenance base cost	0.25%	% of system cost to set Year 1 maint	Amount borrowed	13,959
Maint inflation rate	5.00%	% increase in maint cost per year	Bridge amount	5,982
Inverter replace cost	0.55	per watt, incurred in Year 15	Monthly payment	99.17
Federal tax credit	30%		FNMA 30-year fixed rate (90-day)	3.30%

Making Solar Pay

- **Cash down**: any money that you put down to reduce the size of your loan.

- **Amount borrowed**: This is the base amount of your loan. Some finance companies will loan you the full amount of the system cost but then calculate the loan's amortization and payment schedules based on the assumption that you will apply your 30 percent federal tax credit to the loan balance. The net effect is almost as if they considered only 70 percent of the total loan amount as the amount borrowed. The spreadsheet assumes that arrangement, but you can adjust this input to fit the terms of your loan.

- **Bridge amount**: Assuming the arrangement just mentioned, this is the part of your loan that will be paid off with your tax credit. For example, if your system cost $20,000, it might seem that 70 percent of that ($14,000) is the actual amount borrowed. The rest of the loan ($6,000) is what bridges you between when your PV system is installed and when you pay down your loan with the tax credit. The spreadsheet determines bridge amount by subtracting amount borrowed from system cost. It also accounts for the payoff of the bridge amount in year 1.

Solar Loan Option

The sum of cash down, amount borrowed, and bridge amount should equal the system cost. You can adjust those three inputs to fit your scenario, such as increasing the cash-down amount to reduce what you have to borrow. And if you are not going to use your federal tax credit to pay down your loan balance, the amount borrowed would be the total amount of your loan and the bridge amount would be zero.

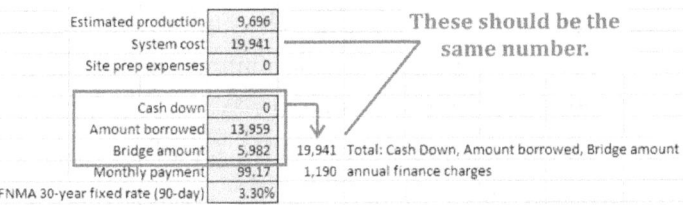

- **Monthly payment**: the amount you pay each month to service the loan, which you should be able to get from the solar proposal or the finance company. The spreadsheet assumes that this payment is fixed, so you will make the same payment each month. That amount is multiplied by twelve to calculate the annual finance charge.

Spreadsheet Columns

The spreadsheet uses those inputs and various calculations to populate data fields for the following columns:

- **Year**: goes from year 0, which is the year that the PV system is installed, to year 25 so that the spreadsheet can give you data for the normal lifespan of a PV system.

- **Production**: shows the annual production of kilowatt-hours. Production for year 1 is the number that you inserted as the estimated production input. Production for year 2 and beyond assumes an annual degradation rate of one half of 1 percent, reached by multiplying the previous year's production number by 0.995. If need be, you can change the system degradation rate in the input section at the top of the spreadsheet.

- **Utility Elec Rate**: shows the rate charged by your utility company for 1 kilowatt-hour of electricity. The year 1 rate comes from what you are currently paying for electricity, as determined in chapter 1 and entered as an input for this spreadsheet. The rates for year 2 and beyond assume an increase each year of 3.5 percent, shown as the default utility-inflation-rate input. In some areas, this escalation will be higher, often going into a range of 5.0 to 6.0 percent per year. You can update that input as needed.

Solar Loan Option

- **System Cost**: the amount that you are paying the solar company to install and commission your PV system.

- **Site Prep Expenses**: any out-of-pocket expenses that you did not pay to the solar company.

- **Cash Down**: As stated in the input section, this is any money that you put down to reduce the size of your loan.

- **Bridge Amount**: the part of the loan that will be paid off with your 30 percent tax credit, assuming your finance company requires you to apply the tax credit to the loan balance in exchange for a lower monthly payment. If your payment is based on the full amount of your loan, the bridge amount should be zero.

- **Loan Balance**: the amount you owe on your loan. For year 0, this is the amount that you borrow. It is then reduced each year as the principal part of your loan payments is applied to the balance.

- **Finance Charges**: the annual amount of your loan payments.

- **Maintenance**: There is very little maintenance with a PV system. The year 1 rate is set as one quarter of one percent, multiplying the system cost by 0.0025. Year 2 and beyond assumes a 5 percent increase over the previous year's maintenance cost.

- **Replace Inverter**: assumes that you will have to replace the inverter around year 15, with the cost being $0.55 per watt. That number is one of the default inputs at the top of the spreadsheet. You might want to check with your solar company for a more accurate number. Also, if you are using microinverters, ask the solar company about the best way to account for those replacement costs.

- **Federal Tax Credit**: This is the 30 percent tax credit, so the spreadsheet adds System Cost and Installation Expenses and then multiplies that number by 0.30.

- **Bridge Payoff**: This is applying your 30 percent tax credit to pay off your bridge amount, if applicable.

- **PBIs & Other Incentives**: the performance-based incentives. Your solar company can identify these for you, if they are available.

Solar Loan Option

- **Annual Elec Savings**: a simple calculation of multiplying production by the utility electrical rate. These savings should rise each year because the rate charged by the utility will likely rise faster than the production level will fall.

- **Net Cash Flow**: gives the financial result for a given year by adding all the revenue and expenses for that time period.

- **Running Total**: shows where you are with your investment by adding the current net cash flow number to the previous year's running total.

- **Future Savings (FS)**: totals the Annual Electric Savings for PV system's remaining productive years. For example, the Future Saving for year 1 is the total of all the annual savings from year 2 through year 25. For year 2, the number is the annual savings from year 3 to year 25, and so on.

- **Estimated Value Add (Present Value of FS)**: uses the current 30-year FNMA fixed interest rate to calculate how much Future Savings (FS) are worth today.

- **Potential Net Impact**: shows the overall financial picture by adding Loan Balance, Running Total, and Estimated Value Add.

What to Look For

If you are financing your PV system with a solar loan, the two numbers to watch are:

- **Net Cash Flow**: tracks the money coming in and going out each year. If you are net positive, the amount that you save each year is greater than the amount you pay to service the loan.

- **Potential Net Impact**: adds together Loan Balance (how much you owe on the loan), Running Total (how much you have saved thus far), and Estimated Value Add (the current value of savings yet to come). A positive number means that the amount saved and potential value added by the PV system are greater than the amount that you owe on the loan.

What conclusions can you draw? First and foremost, you want to be net positive early in the life of the PV system. If you are, then you will have purchased the PV system with little or no money down, the savings produced by the system are covering the cost of the loan, and the loan will be paid off in time.

Second, Potential Net Impact is positive in year 1 and remains positive throughout the life of the PV system. So if you opt to sell your home while you are paying back

Solar Loan Option

that loan, the positive impact of the PV system should be higher than your negative loan balance.

In this example, borrowing the money to purchase a PV system makes financial sense.

The numbers to analyze if you finance your PV system with a loan.

	Loan Balance	Finance Charges	Net Cash Flow	Running Total	Future Savings (FS)	Estimated Value Add (Present Value of FS)	Potential Net Impact
Year 0	-13,959		0	0			
Year 1	-13,561	-1,190	143				5,894
Year 2	-13,142	-1,190	175				6,543
Year 3	-12,702	-1,190	208				7,211
Year 4	-12,235	-1,190	242				7,902
Year 5	-11,739	-1,190	276				8,613
Year 6	-11,223	-1,190	312				9,335
Year 7	-10,678	-1,190	349				10,070
Year 8	-10,099	-1,190	387				10,823
Year 9	-9,485	-1,190	426				11,589
Year 10	-8,843	-1,190	467				12,357
Year 11	-8,159	-1,190	324				12,952
Year 12	-7,440	-1,190	367				13,547
Year 13	-6,672	-1,190	412				14,151
Year 14	-5,870	-1,190	458				14,744
Year 15	-5,018	-1,190	-3,290				11,538
Year 16	-4,125	-1,190	554				12,109
Year 17	-3,176	-1,190	604				12,669
Year 18	-2,171	-1,190	655				13,211
Year 19	-1,117	-1,190	708				13,717
Year 20	0	-1,190	762	4,537	11,356	9,656	14,193
Year 21			2,007	6,544	9,216	8,095	14,639
Year 22			2,065	8,609	7,013	6,363	14,972
Year 23			2,123	10,732	4,744	4,446	15,178
Year 24			2,184	12,916	2,407	2,330	15,246
Year 25			2,246	15,162	0	0	15,162

You are net positive in Year 1.

Potential Net Impact is positive every year, so the accrued savings and added value are greater than the loan balance.

Appendix D
Spreadsheet: Mortgage Option

When to Use This Spreadsheet

This option is where the money needed to purchase the PV system is added to a new mortgage or the refinancing of an existing mortgage. The Mountain Edge Publishing website (www.mepub.com) has spreadsheets for both thirty-year and fifteen-year scenarios.

Keep in mind that your finance company will transfer money to you, which you will then use to pay the solar company. So from the solar company's perspective, it is a cash transaction, which may entitle you to additional discounts related to going with cash.

Spreadsheet Inputs

The Mortgage Option spreadsheet uses the same inputs as the Cash Option spreadsheet (see appendix B) with the addition of three more:

Financial Analysis--Mortgage (30 years)

Input	Value	Description		Output	Value
System size (in kW)	6.90			Estimated production	9,696
Cost per watt	2.89			System Cost	19,941
System degrade rate	0.5%	% loss of production per year		Site prep expenses	0
Base utility elec rate	0.1226	per kWh		Cash down	0
Utility inflation rate	3.50%	% increase in elec rate per year		Added to mortgage	19,941
Maint base cost	0.25%	% of system cost to set Year 1 maint		Added to monthly payment	106.58
Maint inflation rate	5.00%	% increase in maint cost per year			
Inverter replace cost	0.55	per watt, incurred in Year 15			
Federal tax credit	30%			FNMA 30-year fixed rate (90-day)	3.30%

Making Solar Pay

- **Cash down**: any money that you put down to reduce the amount added to your mortgage.

- **Added to mortgage**: the amount your loan balance will increase by adding the PV system's cost into the mortgage. Note: Adding "Cash down" and "Added to mortgage" together should equal the System Cost.

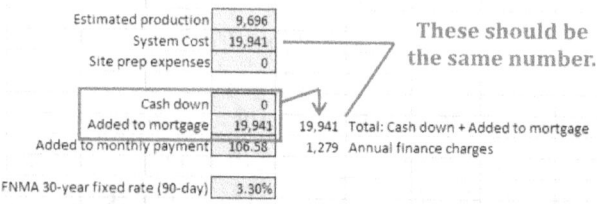

- **Added to monthly payment**: Ask your mortgage company to give you two different payment numbers. First is the amount you would pay each month if you borrowed extra money to pay for solar. Second is the payment each month if you didn't borrow additional funds. The difference between those two numbers is the amount added to your monthly payment. The spreadsheet then multiplies that input by twelve to get the annual finance charge. Note: The spreadsheet assumes the loan rate is fixed, so the annual finance charge remains the same over the life of the loan.

Mortgage Option

Spreadsheet Columns

The spreadsheet uses those inputs and various calculations to populate data fields for the following columns:

- **Year**: goes from year 0, which is the year that the PV system is installed, to year 25 so that the spreadsheet can give you data for the normal lifespan of a PV system.

- **Production**: shows the annual production of kilowatt-hours. Production for year 1 is the number that you inserted as the estimated production input. Production for year 2 and beyond assumes an annual degradation rate of one half of one percent, reached by multiplying the previous year's production number by 0.995. If need be, you can change the system degradation rate in the input section at the top of the spreadsheet.

- **Utility Elec Rate**: shows the rate charged by your utility company for 1 kilowatt-hour of electricity. The year 1 rate comes from what you are currently paying for electricity, as determined in chapter 1 and entered as an input for this spreadsheet. The rates for year 2 and beyond

141

Making Solar Pay

assume an increase each year of 3.5 percent, shown as the default utility-inflation-rate input. In some areas, this escalation will be higher, often going into a range of 5.0 to 6.0 percent per year. You can update that input as needed.

- **System Cost**: the amount that you are paying the solar company to install and commission your PV system.

- **Site Prep Expenses**: any out-of-pocket expenses that you did not pay to the solar company.

- **Cash Down**: as stated in the input section, this is any money that you put down to reduce the amount added to your mortgage.

- **Loan Balance**: the amount you owe on the solar part of your loan. For year 0, this is the amount added to the mortgage. It is then reduced each year as the principal part of your loan payments is applied to the balance.

- **Finance Charges**: the annual solar impact on your loan payments. If you are using the spreadsheet for a fifteen-year mortgage, the spreadsheet will display finance charges from year 1 to year 15. If you have the thirty-year spreadsheet, it will display finance charges for all twenty-five years of the PV system's useful life.

- **Maintenance**: There is very little maintenance with a PV system. The year 1 rate is set as one quarter of one percent, multiplying the system cost by 0.0025. Year 2 and beyond assumes a 5 percent increase over the previous year's maintenance cost.

- **Replace Inverter**: assumes that you will have to replace the inverter around year 15, with the cost being $0.55 per watt. That number is one of the default inputs at the top of the spreadsheet. You might want to check with your solar company for a more accurate number, which you can use to update the inverter-replacement-cost input. Also, if you are using microinverters, ask the solar company about the best way to account for those replacement costs.

- **Federal Tax Credit**: This is the 30 percent tax credit, so the spreadsheet adds System Cost and Installation Expenses and then multiplies that number by 0.30.

- **PBIs & Other Incentives**: the performance-based incentives. Your solar company can identify these for you, if they are available.

- **Annual Elec Savings**: a simple calculation of multiplying production by the utility electrical rate. These savings should rise each year because the rate charged by the utility will likely rise faster than the production level will fall.

- **Net Cash Flow**: gives the financial result for a given year by adding all the revenue and expenses for that time period.

- **Running Total**: shows where you are with your investment by adding the current net cash flow number to the previous year's running total.

- **Future Savings (FS)**: totals the Annual Electric Savings for PV system's remaining productive years. For example, the Future Saving for year 1 is the total of all the annual savings from year 2 through year 25. For year 2, the number is the annual savings from year 3 to year 25, and so on.

- **Estimated Value Add (Present Value of FS)**: uses the current 30-year FNMA fixed interest rate to calculate how much Future Savings (FS) are worth today.

- **Potential Net Impact**: shows the overall financial picture by adding Loan Balance, Running Total, and Estimated Value Add.

Mortgage Option

What to Look For

If you are adding the cost of the PV system to a mortgage, the two numbers to watch are:

- **Net Cash Flow**: tracks the money coming in and going out each year. If you are net positive, the amount that you save each year is greater than the amount you pay to service the loan.

- **Potential Net Impact**: adds together Loan Balance (how much you owe on the solar part of your mortgage), Running Total (how much you have saved thus far), and Estimated Value Add (the current value of savings yet to come). A positive number means that the amount saved and potential value added by the PV system are greater than the amount that you owe on the loan.

What conclusions can you draw? First and foremost, you want to be net positive early in the life of the PV system. You bought the PV system with no money down and the savings produced by the system are covering the additional costs added to your monthly payment. And since this loan is part of your mortgage, the interest payments related to solar will be tax deductible.

Second, Potential Net Impact is positive in year 1 and remains positive throughout the life of the PV system. So if you opt to sell your home while you are paying back

Making Solar Pay

that loan, the positive impact of the PV system should be higher than your negative loan balance.

In this example, including solar as part of your mortgage makes financial sense.

The numbers to analyze if you include solar in your mortgage.

	Loan Balance	Finance Charges	Net Cash Flow	Running Total	Future Savings (FS)	Estimated Value Add (Present Value of FS)	Potential Net Impact
Year 0	-19,941		0	0			
Year 1	-19,642	-1,279	6,036				5,706
Year 2	-19,333	-1,279	86				6,157
Year 3	-19,004	-1,279	119				6,625
Year 4	-18,665	-1,279	153				7,099
Year 5	-18,306	-1,279	188				7,584
Year 6	-17,927	-1,279	223				8,079
Year 7	-17,528	-1,279	260				8,580
Year 8	-17,109	-1,279	298				9,084
Year 9	-16,671	-1,279	337				9,585
Year 10	-16,212	-1,279	378				10,081
Year 11	-15,723	-1,279	235				10,392
Year 12	-15,215	-1,279	278				10,687
Year 13	-14,677	-1,279	323				10,973
Year 14	-14,118	-1,279	369				11,232
Year 15	-13,520	-1,279	-3,379				7,684
Year 16	-12,902	-1,279	465				7,892
Year 17	-12,244	-1,279	515				8,072
Year 18	-11,556	-1,279	566				8,207
Year 19	-10,838	-1,279	619				8,289
Year 20	-10,080	-1,279	673	8,741	11,356	9,656	8,317
Year 21	-9,283	-1,279	728	9,469	9,216	8,095	8,282
Year 22	-8,445	-1,279	786	10,255	7,013	6,363	8,172
Year 23	-7,568	-1,279	844	11,099	4,744	4,446	7,977
Year 24	-6,640	-1,279	905	12,004	2,407	2,330	7,693
Year 25	-5,663	-1,279	967	12,971	0	0	7,308

You are net positive every year except for the year when you replace the inverter.

Potential Net Impact is positive every year, so the accrued savings and added value are greater than the loan balance.

Index

Alternating current (AC) electricity, 23
Analyzing cost, 62-63
Appraisals, 75-78
Array, 21-22
Assumptions, 66
Avoided cost, 71
Battery backup, 34-38
Blackout, 33, 36-37
Buy-all / sell-all system, 114-115
Buying options, 47-56
Cell, 21
Carbon footprint, 57
Case study, 99-106
Cash flow analysis: cash option, 71-74
Cash flow analysis: loan option, 74-75
Cash purchase, 47-49, 53-56, 117-127
Code compliance, 86
Community solar garden, 111-112
Comparing proposals, 57-67
DC direct system, 114
Direct current (DC) electricity, 22
Electric bill, 15, 18-19, 91-92
Electric meter, 23, 93-94
Electrical code compliance, 86
Electrical service panel, 23
Electrical subpanel, 36
Electrical upgrades, 83-84
Electrical usage, 11-17

Fannie Mae 30-year fixed rate, 119-120
Feed-in tariff, 114-115
Financial analysis, 69-82
Future issues, 38-39
Goals, 17-20
Grants, 44
Grid-direct system, 24-33
Grid-direct system with battery backup, 34-37
Historical data, 11-17
Incentives, 41-45
Installation, 83-87
Inverter, 22-23
Kilowatt-hours, 13-14
Leasing, 50-56
Leveling production estimates, 64-65
Loans, 49-50, 53-56, 129-137, 139-146
Local incentives, 44-45
Maintenance, 89-90
Microinverters, 31-33
Module, 21
Monitoring production, 94-98
Net metering, 28
Net positive, 74
Off-grid system, 109-110
Opportunity cost, 48
Optimizers, 30-31
Panels, 21
Partial offset, 18-20
Peak sun hours, 28
Performance monitoring, 91-98
Performance-based incentives (PBI), 44-45, 121
Photovoltaic, 21
Power purchase agreement (PPA), 53

Index

Property tax exemptions, 44
Proposals, 57-67
PV systems, 21-39, 109-115
Rates of return, 48-49, 73-74
Real estate appraisals, 75-78
Rebates, 44
References, 66-67
Renewable Energy Tax Credit, 41-43
Return on investment (ROI), 73-74
Roof issues, 84
Schedules, 67
Service panel, 23
Shade issues, 85
Site audit, 83
Solar panels, 21
Spreadsheet: cash option, 117-127
Spreadsheet: mortgage option, 139-146
Spreadsheet: solar loan option, 119-137
Spreadsheets, 16, 58, 70
Stand-alone system, 109-110
State incentives, 44-45
String inverter, 29-30
Subpanel, 34
Tax credit, 41-43
Tiered billing, 18-19
Types of PV systems, 21-39, 109-115
Value add, 75-78
Variable cost, 18
Warranties, 66
Zero-sell system, 112-113

About the Author

Matt McNearney became interested in solar energy when he explored installing a PV system at his home. He received proposals from five solar companies, but he struggled with how to compare them, especially the financial assumptions and projections. The process that he developed to do that became the framework for this book, the goal being to provide homeowners with an unbiased, systematic way to evaluate solar's potential.

He lives in Denver, Colorado.

www.ingramcontent.com/pod-product-compliance
Lightning Source LLC
Chambersburg PA
CBHW070248190526
45169CB00001B/341